Advanced Praise for
Through a Glass Brightly

"A ruthless and witty debunking of self-flattering illusions held by man over millennia that nonetheless leaves the reader feeling oddly hopeful, and almost giddy. Who knew science could be so much fun?"—Rick Shenkman, author of *Political Animals: How Our Stone-Age Brain Gets in the Way of Smart Politics*.

"A refreshing, revelatory and poignant look at the fundamental faults of our species, that also explains our inability to make the bold decisions ensuring the long-term survival of planet Earth. A must-read for anyone who struggles to comprehend our species and its disregard for the natural world and the impact and consequences of our collective and wasteful existence."—Louise Leakey, Paleontologist and Research Professor, Turkana Basin Institute, Stony Brook University.

"You'd think Copernicus and Darwin would have sufficed to get humanity over its superiority complex, but we are still in the middle of shaking it off. David Barash enlightens us from a solid historical and scientific perspective how far we have come and how far we still have to go."—Frans de Waal, author of *Are We Smart Enough to Know How Smart Animals Are?*

"There could hardly be a more timely and urgent issue than the role of scientific inquiry in determining what makes humans human and our proper place in and relationship to nature. In lucid prose that explains the scientific method to anyone who cares about the difference between facts and fantasy, David Barash explores the psychological, social, and physical perils that are inevitable when human beings regard themselves as being above nature rather than a part of nature. This is a splendid tribute to a human specialness that depends not on having been created by a divine being but on our willingness to use reason to deal wisely with the rest of nature. Every literate politician in Washington should read this book."—Susan Jacoby, author of *The Age of American Unreason in a Culture of Lies.*

"David Barash confronts with friendly erudition and gorgeous range the matter of what is human nature and why humans fight the facts of complicated life so eagerly. He does so with kind verve and a responsible salute to the endless role of science and literature, its mate in seeking meaning."—Lionel Tiger, Charles Darwin Professor of Anthropology Emeritus, Rutgers University, and author, most recently of *The Decline of Males* and *God's Brain.*

"This engaging, energizing and enlightening treatise on man's place in nature goes a long way towards reminding all humanity that we are part of the natural world. But it issues a warning as well: if modern humans continue to ignore this simple fact, it will be at our peril."—Donald C. Johanson, Discoverer of "Lucy" and Founder of the Institute of Human Origins, Arizona State University.

Through a Glass Brightly

Using Science to See Our Species as We Really Are

David P. Barash

OXFORD
UNIVERSITY PRESS

OXFORD
UNIVERSITY PRESS

Oxford University Press is a department of the University of Oxford. It furthers
the University's objective of excellence in research, scholarship, and education
by publishing worldwide. Oxford is a registered trade mark of Oxford University
Press in the UK and certain other countries.

Published in the United States of America by Oxford University Press
198 Madison Avenue, New York, NY 10016, United States of America.

Library of Congress Cataloging-in-Publication Data
Names: Barash, David P., author.
Title: Through a glass brightly : using science to see our species as we really are / David P. Barash.
Description: New York, NY : Oxford University Press, [2018] |
Includes bibliographical references and index.
Identifiers: LCCN 2017043946 | ISBN 9780190673710
Subjects: LCSH: Human beings—Animal nature. | Philosophical anthropology. |
Human ecology.
Classification: LCC GN280.7 .B37 2018 | DDC 599.9—dc23
LC record available at https://lccn.loc.gov/2017043946

3 5 7 9 8 6 4 2

Printed by Sheridan Books, Inc, United States of America

Contents

Part I: The Allure of Human Centrality, or, How We Persistently Try to Deny Our Place in the Natural World—and Fail

Prelude to Part I 2

1. The journey to Brobdingnag 16

2. From centrality to periphery 19

3. The meaning of life 25

4. Well designed? 33

5. The anthropic principle 39

6. Tardigrades, trisolarans, and the toughness of life 52

7. Of humanzees and chimphumans 63

8. Separateness of self? 68

Part II: New Ways of Understanding Human Nature

Prelude to Part II 84

9. Uniquely thoughtful 93

10. Conflict between parents and offspring 111

11. True or false? 125

12. The myth of monogamy 135

13. War and peace 146

14. About those better angels 162

15. Who's in charge? 172

16. The paradox of power 179

Conclusion: *Optāre aude* 190

Index 195

Through a Glass Brightly

Part I
The Allure of Human Centrality, or, How We Persistently Try to Deny Our Place in the Natural World—and Fail

Prelude to Part I

BEING KICKED OUT of paradise must have been tough on Adam and Eve. The requirement to earn food by the sweat of their brow was doubtless bad enough, not to mention pain in childbirth, but losing immortality and being told that dust they are and to dust will they return, must have been—in modern parlance—a real bummer.[a]

In any event, those particular losses happened long ago, so no one around today has directly experienced the trauma. By contrast, in recent time we human beings have been deprived of some of our most beloved and comforting Edenic myths, with others dropping out almost daily, leaving us to confront a growing array of paradigms lost, many of them variants on this fact, easy to state, hard to accept: we are not as important, as special, as all-around wonderful as we'd like.

Confronting this reality—more accurately, trying to negotiate around it—engages a deep-rooted tendency whereby people begrudgingly accept what is forced on them, while nonetheless clinging to their most cherished preexisting beliefs, what they *want* to be true. Prominent current cases include granting that the Earth's climate is heating up but refusing to accept that human beings are responsible, or acknowledging that evolution is real when it comes, for example, to microbial antibiotic resistance but denying that it produced us.

It is an inclination that is evidently deep-rooted in the human psyche. Thucydides, fifth-century BC historian of the Peloponnesian War, complained of his contemporaries that "their judgment was based more upon blind wishing than upon any sound prevision; for it is a habit of mankind to entrust to careless hope what they long for, and to use sovereign reason to thrust aside what they do not fancy."

It is difficult to modify an opinion once established, especially if it's a favorable one—and even more so if it is centered on one's self (or one's species). A particular consequence of intellectual progress has nonetheless been an understanding of our increasingly deflated place in nature and in the universe, making it more and more

---

[a] As a biologist, I confess to an added personal regret that the serpent was punished by being forced to crawl on its belly, since I would have loved to see how it might have ambulated otherwise: perhaps bouncing on its tail as on a pogo stick.

untenable to see ourselves as somehow outside of—never mind superior to—the rest of "creation." *Through a Glass Brightly* therefore outlines a less grandiose but more bracingly accurate view of ourselves, thanks to modern science.

It is the height of paradox. The more we learn about our own species and the world around us, the more we are humbled, forced to relinquish some of our most cherished illusions, especially our unique centrality. This demotion was given a dramatic push by Copernicus, Kepler, and Galileo, who upended the Ptolemaic view that the cosmos revolves around a central and immobile planet Earth. It is difficult for us in the twenty-first century to appreciate how troublesome, even painful, it was for our home—and by extension, ourselves as well—to be so irrefutably downgraded.[b] Emblematic of this, the sixteenth-century Danish astronomer Tycho Brahe (one of the finest scientists of his day and probably the greatest naked-eye astronomer of all time) proposed an alternative to the Copernican system. According to Brahe's paradigm, the five known planets—Mercury, Venus, Mars, Jupiter, and Saturn—all circled the Sun, but that conglomeration in turn revolved around a central Earth. Many astronomers note, incidentally, that Brahe's proposed system was in fact a good fit with the data available to him, and that his "blunder" wasn't so much a result of prevailing religious belief as an understandable reluctance to discard the reigning Earth-centered system and replace it with the newer solar one unless the evidence was indisputable.

Such an adjustment was, however, ultimately necessary, and although immensely consequential, it was part of an even broader and deeper substitution, what the pioneering sociologist Max Weber called the "disenchantment of the world," exemplified by Galileo's more general discovery that the world lends itself to material explanations: objective as distinct from subjective, natural rather than supernatural. In *The Myth of the Machine*, the historian Lewis Mumford complained,

> Galileo committed a crime far greater than what any dignitary of the Church accused him of; for his real crime was trading the totality of human experience for that minute portion which can be observed and interpreted in terms of mass and motion . . . . In dismissing human subjectivity Galileo had excommunicated history's central subject, multi-dimensional man . . . . Under the new scientific dispensation . . . all living forms must be brought into harmony with the mechanical world picture by being melted down, so to say, molded anew to conform to a more mechanical model.[1]

---

[b] It has been argued, by the way, that not all contemporary theologians and philosophers felt that the center of the universe was such a good place to be. Thus, the center of the Earth was widely considered to be the abode of hell, and the center of the universe, not much better.

It is said that after being forced to recant his claim that the Earth moves around the sun, Galileo muttered to himself, "E pur si muove" (And yet it moves). The story might be apocryphal, but the underlying mechanical model, the cosmic machine of which everyone and everything is a part, is no myth. Nor is the resistance that it continues to evoke.

After the Copernican revolution and the one that Galileo initiated (which is still a work in progress) came the Darwinian revelation that we, along with the rest of the living world, aren't the products of Special Creation, but rather the results of a natural, material process that physically connects us to all other organisms. Even now, opponents of evolution cling desperately to the illusion that human beings—and, in some cases, living things generally—are so special that only a benevolent Creator could have produced them. For these people, it remains a hard sell that the organic world, like the sun and its five planets, doesn't revolve around us.

The third major leg of this troublesome triad was initiated by Freud, who (despite his occasional crackpot flights of fancy) came up with at least one solid and highly consequential discovery: the existence of the unconscious. Regardless of what one thinks of "penis envy," the "Oedipus complex," and so forth, there is general agreement that the human mind is like an iceberg, with much of its mass hiding below the conscious waterline.

So, not only have we been kicked out of our presumed astronomical centrality, immersed in a world of materiality and deprived of our widely assumed creaturely uniqueness, but we aren't even masters in what seemed to be left to us, our pride and joy: our rational, conscious minds.

Of course, there are many people for whom the more we learn about the natural world, the more wonderful it is revealed to be, and thus, the more magnificent its Creator. It is likely, nonetheless, that insofar as human beings are perceived as "natural," and thus explicable in terms of widely accepted scientific principles rather than uniquely fashioned by supernatural intervention, the more resistance will be evoked among those committed not just to human specialness but also to perceiving this specialness as evidence for divine power and intervention. It is hard enough to adjust your opinion—think of how much easier it is to change your clothes than to change your mind—harder yet to relinquish a cherished perspective. Especially one that has the blessing of religious belief. As Jonathan Swift noted centuries ago in his essay, *Seeking Wisdom*, "You cannot reason a person out of a position he did not reason himself into in the first place."

The only constant, nevertheless, is change. The story is told of an ancient ruler who tasked his advisers to come up with a statement that would be true at all times and for all occasions. Their response: "This too shall pass." But although the world's factual details are constantly shifting (as the philosopher Heraclitus pointed out, you cannot step in the same river twice, and, as Buddhists note, all things are impermanent), the basic rules and patterns underlying these changes in the physical and biological world are themselves constant. So far as we know, light traveled at the same speed during the age of dinosaurs, during the Peloponnesian War, and today. The Second Law of Thermodynamics is true and was true long before Carnot discovered this principle, just as special and general relativity was valid before being identified by Einstein.

Compared to the apparently unchanging nature of physical law, our insights are always "evolving," along with living things themselves, although recognizing and understanding these insights often requires a major paradigm shift. Interestingly, although much has been learned (and more yet, hypothesized!) about how science proceeds to generate reliable knowledge, relatively little is known about how and why people—including scientists themselves—change their personal beliefs. On the one hand, we have Max Planck's famous quip, "A new scientific truth does not triumph by convincing its opponents and making them see the light, but rather because its opponents eventually die, and a new generation grows up that is familiar with it." And on the other, the more optimistic and probably widespread notion that, eventually, the truth will out.

To be clear, I am not claiming that clinging to factual error is necessarily the result of benighted religious prejudice or the simple psychology of denial. Sometimes, incorrect scientific ideas enjoy popularity because they are a good fit with current empirical data. Initially, nearly all competent astronomers resisted Copernicus's model, at least in part because it didn't accord any better with astronomic observations than did the regnant Ptolemaic one. However, at least some of that resistance was due, as well, to the painful emotional and theological reorientation necessitated by its acceptance.

"All truths are easy to understand once they are discovered," wrote Galileo. "The point is to discover them."[2] Much as I revere Galileo, I am not at all sure that in this regard he was correct. Sometimes, the problem isn't simply to *discover* truths but to *accept* them, which is especially difficult when such acceptance requires overcoming the bias of anthropocentrism, whereby people put their own species at the center of things. Although my hope is that seeing *Homo sapiens* through the bright glass of science will contribute to humanity understanding and accepting itself, given the stubborn persistence of anthropocentric thinking, I cannot promise success. The writings of the "new atheists" offer a possible parallel: Dawkins, Harris, Dennett, and Hitchens do not appear to have converted people to atheism so much as they have helped initiate a discussion, such that even though atheism is not nearly (yet?) mainstream, it has become more respectable.

Thomas Kuhn famously suggested that each science operates within its own paradigm, which limits the ability of its practitioners to conceive other approaches—until a "paradigmatic revolution" supplants the prior intellectual system with a new one, which in turn, is similarly limiting. A related problem is that of "unconceived alternatives," whereby our ability to make sense of natural phenomena is restricted by our own failures of imagination. (After all, an evolutionary perspective suggests that the human mind has evolved to maximize the fitness of its possessor, not necessarily to provide accurate information about the world.) To this must be added a seemingly focused resistance to what I label anthropodiminution, whereby alternatives that demote humanity's already wounded self-image are especially hard to conceive.

The present book is intended to help undercut this resistance, in part by emphasizing that our naturalness does not really diminish us, except with respect to the notion that human distinctiveness derives from being in some sense cut off from the rest of the natural world. It contends, by contrast, that pointing out humanity's materiality—and thus, our profound linkage to everything else—only enlarges and thereby enhances our status.

In his *Ethics*, Spinoza wrote, "men commonly suppose that all natural things act, as men do, on account of an end [and] . . . that God has made all things for man, and man that he might worship God." Spinoza felt that this belief was nonsense: "The wise man seeks to understand Nature, not gape at it like a fool."[3] And definitely not to see Nature as serving human ends or to perceive ourselves as central to Nature, which itself must be subject to human concerns.

This puts certain popular practices in a different light. For example, the craze to take "selfies," thereby inserting one's image into social situations and natural scenes, may well be powered by the satisfaction of being literally incorporated into a circumstance for which one is otherwise peripheral—if present at all: "Look at me! Here I am!" Similarly with the use of the Internet as a self-reinforcing echo chamber, whereby people arrange to encounter ideas and perspectives that accord with their own, thus not only feeding their particular interests but also (more to the point and more unfortunately) reinforcing the illusion that their ideas and perspectives are central to public discourse such that alternative views hardly exist. And that, in any event, they don't count for very much.

A perspective based on science is a corrective to this seductive myopia, whereby we are all tempted to see ourselves as bigger, more important, more central to the world than we really are. Admittedly, sometimes it helps to be in over our heads, insofar as this helps us understand just how tall we really are. Sometimes, accordingly, it is in our interest that as Browning suggested, our reach exceeds our grasp, since we expand ourselves and our capabilities by striving, by "demanding the impossible" instead of settling for whatever is readily available and achieved without effort. But although effort is good, delusions are not, and misconceptions of grandeur and of centrality are not only inaccurate but potentially dangerous. There is no better corrective to such delusions than the cold shower of scientific inquiry. Paradigms lost can thus be wisdom gained.

IT IS EASY to think of science as essentially organized common sense, based as it is on generating hypotheses, testing them, evaluating the results of those tests, and then, if the findings are consistent (especially if they are coherent with a prior, unified body of theory) and if the predictions aren't falsified over time, concluding that the results are scientifically meaningful, whereupon they are added to our body of knowledge. Science is a phenomenally powerful tool, the strongest and most effective yet devised. Contrary to widespread assumptions, however, science is at its most useful when its specific findings go counter to common sense. Indeed, science can usefully be conceived as a *corrective* to it. Otherwise, we wouldn't need science; we could simply "go with our gut."

Isaac Asimov (who was a highly regarded biochemist before he became a famous author) once noted science is "a way of trying to improve your knowledge of nature, it's a system for testing your thoughts against the universe and seeing whether they match."[4] Often they don't, and when that happens, it isn't the universe that is wrong.

Intuition can be a misleading guide, even when it comes to something as seemingly cut and dried as physics. For example, it is tempting to assume—as did notable thinkers since Aristotle—that a heavy object would fall more rapidly than a light one. This was widely taken as a commonsensical "fact" until Galileo demonstrated that it isn't true (although there is some doubt whether, as widely thought, he actually tested this by dropping two objects from the leaning tower of Pisa). Or take a ball on a string and swing it around in a circle. Now ask yourself: if you let it go while it is rotating, what path will it take? Many people—even college-educated science majors—assume that it would travel in a spiral. But it won't. It will continue in a straight-line tangent to the circular route that it had been following.

There are many other cases in which what seems obvious is wrong. The sun doesn't go around the Earth, as it appears to do. That same Earth isn't flat, as it seems to be. Apparently solid objects are actually composed of mostly empty space. Science is a pushback against the errors that are frequently incorporated into what is often taken for granted. It is closer to the Enlightenment than to Romanticism, basing its insights on skeptical inquiry, data, analysis, interpretation, and debate rather than gut feelings. It takes, after all, an outright denial of intuition to acknowledge that tiny organisms— much smaller than anything we can see with the unaided eye—can make us ill. Hence it is disturbingly easy for the antivaxer movement to gain adherents, even though being unvaccinated is immensely more dangerous than the alternative.

This route leads, of course, to our own species and the pervasive common sense perception that *Homo sapiens* is discontinuous from the rest of nature, and that even aside from the presumption that we have souls and "they" do not, human beings are more advanced than other animals—once again presumably because of our big brain and what it can do. The opposite of "advanced" is primitive, as in "a flatworm has a brain which, compared to that of a person, is primitive." In her book, *The Incredible Unlikeliness of Being: Evolution and the Making of Us*, Alice Roberts[5] points out, "All tetrapods around today have five digits or fewer at the end of their limbs. So it seems reasonable to assume that we've all descended from a five-fingered, or pentadactyl, ancestor." Accordingly, at least with respect to our toes and fingers, we are primitive rather than advanced. On the other hoof, by contrast, a horse's toe, at the end of each limb, consists of but a single digit, making it more advanced than ourselves, at least when it comes to its tip-toe middle digit—technically the digitus impudicus—whereby *Equus caballus* is more different from the ancestral vertebrate condition than we are. (This also means, Dr. Roberts notes, that horses are walking around giving us the finger.)

In most other respects, our demotion—more accurately, our inclusion in the material bestiary of the real world—courtesy of science, is not only long overdue but somewhat more serious. When Carl Sagan famously informed his television audience that we are all made of "star stuff," the deeper implications may well have been lost on many of his fellow star-stuffed critters. Please meditate, for a moment, on the fact that there is literally nothing special about the atoms of which everyone is composed. Even in their statistical preponderance by mass, these elements reflect rather well the chemical composition of the universe as a whole: oxygen, carbon, hydrogen, nitrogen, calcium, and so forth. Of course, there *is* something special about the way these common components are arranged; that's the work of natural selection, which, when presented with alternatives, multiplied and extended the frequency of those combinations that were comparatively successful in replicating themselves. All this, in turn, further highlights the degree to which we are cut from the same cloth.

Recall Socrates's dictum, "The unexamined life is not worth living." The issue, with respect to the present book, is not so much examining your own life, or human life generally, but rather, understanding both and doing so with humility, honesty, and an expanded sense of interconnectedness and potential. According to the King James Version of the Bible, in 1 Corinthians 13:12, Paul wrote, "For now we see through a glass, darkly," an observation that—suitably modified—led to the title of the present book. Paul went on to write that after this restricted, darkened field of vision, we could look forward, upon meeting God, to seeing "face to face," adding, "now I know in part; but then shall I know even as also I am known." Fine for believers, but for the secularists among us, there is even better news: through the glass of science, we can all know and be known, and see brightly, not in heaven but here and now.

Yet there is some wisdom in Paul's "darkly," namely that we don't necessarily see the world with perfect accuracy. Why not? Because we haven't evolved to do so. The fact that we can penetrate some of the universe's deeper secrets, unravel our own DNA, and so forth, is remarkable, but not literally miraculous. Just as the human nose didn't evolve to hold up eyeglasses, but does a good job at it, and binocular vision evolved to enable our arboreal primate ancestors to navigate their three-dimensional lives and has subsequently done a good job enabling us to throw objects accurately, drive cars, and pilot airplanes, our five senses along with our cognitive complexity and sophistication evolved for many possible reasons, including navigating an increasingly complex and sophisticated social life, engaging in elaborate communication skills, making and manipulating tools and other devices, predicting the future, and so forth.

Once it became part of our armamentarium, human intelligence and perception has underwritten all sorts of additional activities, such as exploring the universe as well as our own genome and composing symphonies and epic poems; the list is nearly endless, but the basic point is that we didn't evolve with an explicit adaptive capacity to do these things. They were repurposed from neuronal structures and capabilities that emerged for other reasons, not unlike pedestrian curb cuts that have been engineered to permit wheelchair access from street to sidewalk, but are now used at least as much by bicyclists and skateboarders. The biological reality is that our perceived separateness may well have evolved so as to promote the success of our constituent genes, but at the same time, there was little or no evolutionary payoff in recognizing not so much our limitations as our lack thereof.

John Milton wrote *Paradise Lost* to "justify God's ways to man." In the end, what justifies science to men and women is something more valuable and, yes, even more poetic than Milton's masterpiece or Paul's vision: the opportunity to consume the fruits of our own continually reevaluated, deeply rooted, admittedly imperfect, and yet profoundly nourishing Tree of Scientific Knowledge, whereby we increasingly understand ourselves as we really are. I hope that most people will find more pleasure than pain in using science to do so, and in the process, seeing themselves and their species more accurately and honestly—more brightly, in every sense of that word—than ever before.

Since this hope might well seem overly optimistic—even downright smug—this is a good time to introduce something of a counternarrative, a brief meditation on Piss-Poor Paradigms Past: examples of received wisdom that, in their time, went pretty much unquestioned, even among those constituting the scientific establishment. My purpose here is not to cast doubt or aspersions on the scientific enterprise. Quite the opposite. It is to remind the reader that science is an ongoing process, and that whereas the Tree of Scientific Knowledge is a many splendored thing, it also consists of many branches that have ultimately proven to be weak—in some cases, perilously so.

Ironically, some people lose faith in science because of the regular revisions it undergoes, the irony being that it is precisely because science is constantly being revised that we are getting closer and closer to what we can unblushingly call the truth. In short, "what makes science right is the enduring capacity to admit we are wrong."[6]

And there is no doubt that wrong has happened; science, or what used to pass for science, has undergone much pruning, in the course of which the following limbs (once thought strong) are among the many that have been amputated: vitalism (the idea that living things possess some sort of unique life force or "élan vital"), spontaneous generation (rats and maggots emerge from garbage, etc.), confidence that alchemy would enable its practitioners to turn base metals into gold, and widespread and stubborn belief in weird substances such as luminiferous aether, phlogiston, and caloric.

In retrospect, these now discredited concepts, which seem downright foolish via 20-20 hindsight, were reasonable in their day. Take the aether, which was seen as necessary to understand the otherwise mysterious behavior of light. So clear-cut was its apparent legitimacy that James Clerk Maxwell—probably the greatest physicist of the nineteenth century, and whose equations for electromagnetism are still fundamental today—asserted that of all theoretical concepts in physics, the aether was the most securely confirmed. In agreement were two of Maxwell's most notable contemporary physicists: Lord Kelvin and Heinrich Hertz. The latter's research on the propagation of radio waves had given further credence to the consensus that aether was needed as a substance through which both light and radio waves were transmitted.

For centuries, scientists also assumed the dogma of an unchanging Earth and a solid-state universe—now dramatically replaced by continental drift and the Big Bang, respectively. Britain's renowned astronomer-royal Fred Hoyle coined the phrase "Big Bang" as a sarcastic response to what he perceived as a ludicrous alternative to the then-regnant concept of an unchanging cosmos. Now the Big Bang is received wisdom, along with the finding that there are signs of prior water on Mars, but no artificial canals, the existence of which was claimed by Percival Lowell, another famous astronomer.

Some of the most dramatic scientific paradigm shifts have involved biomedicine. Consider, for example, the long-standing insistence that there are four humors—blood, yellow bile, black bile, and phlegm, corresponding, it was thought, to human temperaments: sanguine, choleric, melancholic, and (no surprise here) phlegmatic, respectively. And don't forget bloodletting as a widely acknowledged and scientifically "proven" medical treatment, now known to have hastened George Washington's death and long practiced through much of the Western world. (The term "leech," historically applied to physicians, didn't derive from their presumed avariciousness, but rather, from the use of blood-sucking leeches as an instrument for ostensibly therapeutic exsanguination.)

Thanks to Pasteur, Koch, Lister, and other pioneering microbiologists, we have come to understand the role of pathogens in causing disease, resulting in the scientific discovery that "germs are bad." This particular paradigm—displacing belief in "bad air" and the like ("influenza" derives from the supposed "influence" of miasmas in causing disease)—was vigorously resisted by the medical establishment. Doctors who would routinely go from conducting autopsies on disease-ridden corpses couldn't abide the idea that their unwashed hands were transmitting illness to their obstetric patients, to the extent that Ignaz Semmelweis, who demonstrated the role of hand-borne pathogens in causing "puerperal fever," was ignored, then vilified, even, it appears, literally driven mad.

More recently, however, just as people have finally adjusted to worrying about creatures so small that they can't be seen by the unaided eye, a new generation of microbiologists have demonstrated the stunning fact that most microbes (e.g., including but not limited to the gut microbiome) aren't merely benign but essential for health. Nerve cells, we were long told, didn't regenerate, especially not within the brain. Now we know that actually they do. Brains can even produce whole new neurons; you *can* teach old dogs new tricks.

Similarly, it was assumed until recently that once an embryonic cell differentiates into, say, a skin or liver cell, its fate is sealed. The advent of cloning technology has changed this, with the finding that cell nuclei can be induced to differentiate into other tissue types. Dolly the sheep was cloned from the nucleus of a fully differentiated mammary cell, proof that the paradigm of irreversible cell differentiation itself needed to be reversed, especially—we now know—in the case of embryonic stem cells.

Until recently, physicians were scientifically certain that at least a week of bed rest was necessary after even a normal, uncomplicated vaginal childbirth, not to mention invasive surgery. Now surgical patients are typically encouraged to walk as soon as possible. For decades, protuberant but basically benign tonsils were unceremoniously yanked whenever a child had a sore throat. Not any more. Psychiatry offers its own pervasive, problematic panoply of paradigms past (and good riddance to them!). Until 1974, homosexuality for example was considered a form of mental illness, schizophrenia was thought to be caused by the verbal and emotional malfeasance of "schizophrenogenic mothers," and prefrontal lobotomies were the scientifically approved treatment of choice for schizophrenia, bipolar disease, psychotic depression, and sometimes, merely a way of calming an ornery and intransient patient.

The catalog is extensive. Despite the claim that we have reached the "end of science," the reality is otherwise. For decades, the best scientific advice asserted, for example, that gastric ulcers were produced by stress, especially the hyperresponsiveness of people with "Type A" personalities. Then, in the 1980s, the Australian scientists Barry Marshall and Robin Warren demonstrated that most gastric ulcers are produced by a particular bacterium, *Helicobacter pylori*. In recognition of their paradigm-busting discovery, which had been vigorously resisted by the most scientifically esteemed minds in gastroenterology, Marshall and Warren received a Nobel Prize in 2005. There isn't time and space in the present book to explore the on again–off again controversies over the health consequence of dietary cholesterol, red wine, caffeine, and so forth. A cartoon in *The New Yorker* showed a large, glowering, shapeless something-or-other poised outside a bakery, with the caption reading "The Gluten's back. And it's pissed."

THROUGH A GLASS BRIGHTLY is divided into two parts:

1. Major paradigm shifts that involve diminishment of humanity's self-image, and which have therefore been resisted with particular vigor, for example, heliocentric astronomy, the notion that human beings have been especially "well designed," and so forth.

2. Reassessments of certain previously held notions that deal with specific aspects of "human nature," many of them still alive in the public mind; for example, altruism cannot be explained by evolution and must therefore be a gift from god, and people are naturally monogamous. Here, my intent is less to argue against human central- ity per se than to take issue with an array of preexisting beliefs that have themselves been privileged at least partly because they place human beings in a flattering but misleading light.

Accordingly, whereas Part I looks at (and seeks to debunk) the underlying concept of human centrality, Part II examines "human nature" in greater detail, showing that here, too, we find ourselves less special and more "natural" than an anthropocentric perspective on the human condition would like. In *Anti-Semite and Jew*, Jean-Paul Sartre wrote that the underlying basis of existential freedom can be found in what he calls "authenticity," the courage and capacity to have "a true and lucid consciousness of the situation, in assuming the responsibilities and risks it involves, in accepting it in pride or humiliation, sometimes in horror and hate."[7]

Lest there be any misunderstanding, I am not a species-hating human being, although I maintain that we—along with the rest of the planet and its inhabitants— would all be better off if our species-wide narcissism were taken down a peg or two, if we were to perceive ourselves with less *Homo*-centric delusions of grandeur. Science is supposed to be divorced from pride, humiliation, horror, and hate, and to a large extent, it is. However, as the biological anthropologist Matt Cartmill pointed out in a brilliant essay more than 25 years ago,[8] when it comes to scientific investigations into humanness, there has been a persistent tendency to move the goal posts when- ever other species turn out to have traits that had previously been reserved for *Homo sapiens* alone. As soon as our biological uniqueness is challenged, there has been a scramble to redefine the characteristic in question so as to retain precisely that specialness.

Take brain size. Intelligence is obviously one of our most notable characteristics, which led to the assumption that the human brain must be uniquely, extraordinarily, exceptionally, and altogether wonderfully large. But as Cartmill pointed out, the weight of the *Homo sapiens* brain (1–2 kg) bumped up against the awkward fact that the brains of elephants are larger (5–6 kg), and those of whales (up to 7 kg) are larger yet. This unwanted and uncomfortable reality brought forth a focus on *relative* brain size—comparing species by looking at brain weight in proportion to body weight. Gratifyingly, it happens that this number is substantially higher for *Homo sapiens* (1.6%–3.0%) than for elephants (0.09%) or whales (0.01%–1.16%). So far, so good.

Cartmill noted, however, that even in the realm of relative brain size, we are equaled or exceeded by that of many small mammals, including squirrel monkeys (2.8%–4.0%), red squirrels (2.0%–2.5%), chipmunks (3.0%–3.7%), and jumping mice (3.4%–3.6%). And so, "algometric analysis" was then "invoked to rescue the axiom of human cerebral preeminence. The first step in such an analysis is to assume that the interspecific regression of the logarithm of brain weight on that of body weight ought to be a straight line." Without getting into the details of algometric analysis, suffice it to say that even with this mathematical adjustment, porpoises ended up being "embarrassingly" close to human beings and so another way out was needed. What about assuming that brain size should be proportional to an organism's total metabolic energy expenditure, that is, looking at the amount of energy invested in each creature's brain in proportion to its total energy budget? Sure enough, if we obtain a measure of total metabolic expenditure, by multiplying body weight times baseline metabolic rate, it turns out that porpoises invest proportionately less energy in brain maintenance than do human beings. Even in this case, however, there is a problem, since as Cartmill observed, it is "a maneuver that a lizard might with equal justice use to prove that mammals don't really have bigger brains than reptiles, but only higher metabolic rates."

The above brain brouhaha doesn't even touch the case of learning capacities among insects, whose brains are small indeed: fruit flies average only about 250,000 neurons per brain, and yet they are capable of learning to avoid certain stimuli and to seek out others, to orient themselves via a mental map of their surroundings, and so forth. Moreover, bumblebees—which have approximately 1 million neurons in their brains (a gratifyingly small number compared to mammals)—have recently been shown capable of learning to do something unlike any behavior they are likely to encounter in nature, namely to roll a little ball into the center of a platform in order to receive a small dose of sugar water. Not only that, but individual bumblebees also learn this relatively complex and heretofore unfamiliar behavior more rapidly if given the opportunity to watch other bees learning the task.[9] "Observational learning" of this sort had previously been considered a sign of higher mental powers, especially found in, well, us.

Writing about shared "intellectual faculties," Darwin conceded in his 1871 book, *The Descent of Man, and Selection in Relation to Sex*, "Undoubtedly, it would have been very interesting to have traced the development of each separate faculty from the state in which it exists in the lower animals to that in which it exists in man; but neither my ability nor knowledge permit the attempt." A lot has happened in the intervening time, and although the evidence is accumulating rapidly, it is also resisted by many—and not just religious fundamentalists and spokespeople for the beef and dairy industries.

The struggle against recognizing mental continuity between humans and other animals has taken place in many domains, including, for example, language, the meaning of which has regularly been revised whenever detailed research revealed that nonhuman animals possessed it. Once it became evident that other creatures communicated sophisticated information to each other (such as the "dance of the bees," whereby a forager communicates complex information about the location and even the desirability of a food source to her hive-mates) language was redefined as synonymous with something else: the establishment of arbitrary signs, such as the word "dance" meaning a pattern of complex, rhythmic movements, as opposed to whatever is involved in doing any particular kind of dance.

The influential cultural anthropologist Leslie White spoke for many when he asserted,

> The lower animals may receive new values, may acquire new meanings, but they cannot create and bestow them. Only man can do this . . . . And this difference is one of kind, not of degree . . . . Because human behavior is symbol behavior and since the behavior of infra-human species is non-symbolic, it follows that we can learn nothing about human behavior from observations upon or experiments with the lower animals.[10]

Then, it having been demonstrated that some animals are in fact capable of assigning meaning to arbitrary signals, a movement began to identify not signs, but syntax as the *sine qua non* of "real" language—which is to say, what human beings do.

Note the distinct echoes of Tycho Brahe, developing new and creative ways to retain humanity's special place. The persistent search for human exceptionalism whereby our biology renders us discontinuous from other animals is, if not quite a fool's errand, one persistently undertaken by a subset of *Homo sapiens* who—so long as they base their search on science rather than metaphysics or theology—are doomed to disappointment.

The best view in Warsaw, Poland, is from the top of the Palace of Science and Culture, because that is the only place in the city from which one cannot see this example of Stalinist architecture at its worst. Being too close to the object of our scrutiny is inevitably a problem, which makes it all the more difficult—as well as important—to take a close and careful look at ourselves, mindful that any such view (even, perhaps, the evolutionary one promoted in this book) is liable to distortion.

Nonetheless, as T. S. Eliot proclaimed,[11] and in which I suggest substituting "ourselves" for "the place":

> We shall not cease from exploration,
> and the end of all our exploring
> will be to arrive where we started
> and know the place for the first time.

Add to this, as well: "and not for the last time, either."

NOTES

1. Lewis Mumford, *The Myth of the Machine* (New York: Harcourt, 1967).
2. Galileo, "*Dialogue Concerning the Two Chief World Systems*," 1632.
3. Spinoza, *Ethics*.
4. Asimov cited in Maria Popova, "Isaac Asimov on Science and Creativity in Education," https://www.brainpickings.org/2011/01/28/isaac-asimov-creativity-education-science/.
5. Alice Roberts, *The Incredible Unlikeliness of Being: Evolution and the Making of Us* (London: Heron Books, 2016).
6. L. Rosenbaum. "The March of Science—the True Story," *The New England Journal of Medicine* 377, 2 (2017): 188–191
7. Jean-Paul Sartre, *Anti-Semite and Jew*. (New York: Grove Press, 1962).
8. M. Cartmill, "Human Uniqueness and the Theoretical Content in Paleoanthropology," *International Journal of Primatology* 11 (1990): 173–192.
9. O. J. Loukola, C. J. Perry, L. Coscos, and L. Chittka. Bumblebees Show Cognitive Flexibility by Improving on an Observed Complex Behavior. *Science* 135 (2017): 833–836.
10. Leslie White. *The Science of Culture* (New York: Grove Press, 1949).
11. T. S. Eliot. *Four Quartets* (New York: Harcourt, 1943).

1

The Journey to Brobdingnag

GULLIVER'S TRAVELS—ESPECIALLY the eponymous narrator's adventures in the land of Lilliput, where he was a giant among miniaturized people—is largely read these days (insofar as it is still read at all) as a children's fantasy. This is a shame, because Jonathan Swift's eighteenth-century novel is a biting adult satire, including, for example, an absurd war between the Lilliputians and their rivals, the Blefuscudians, over whether soft-boiled eggs should be opened at their large or small ends.

Swift intended Gulliver to be a kind of everyman, and in a sense our species-wide journey via science has made us all into Lilliputians, shrinking ourselves even as our knowledge has expanded. It is therefore appropriate that on a subsequent voyage, Gulliver met the Brobdingnagians, giants to whom he was tiny and insignificant, mimicking our cosmic situation. The inhabitants of Brobdingnag were not only physically huge, but morally better than the tiny but "full-sized" Gulliver. In fact, when Gulliver attempts to enlighten the Brobdingnagian king as to the virtues of the English, he is admonished that history is not as Gulliver describes; rather, it is a "heap of conspiracies, rebellions, murders, massacres, revolutions, banishments, the very worst effects that avarice, faction, hypocrisy, perfidiousness, cruelty, rage, madness, hatred, envy, lust, malice, or ambition could produce." The king concludes with the oft-quoted observation that people in general and the English in particular are "the most pernicious race of little odious vermin that nature ever suffered to crawl upon the surface of the earth." In short, we are genuinely small, not just in stature compared to the enormous inhabitants of Brobdingnag, but in moral terms as well.

Swift was, admittedly, no great fan of human beings, writing in a letter to the poet Alexander Pope, "I hate and detest that animal called man, although I heartily love John Peter, Thomas and so forth." In Gulliver's voyage to Brobdingnag, human beings are presented, not surprisingly, as downright inconsequential. This is not simply because in the land of giants the diminutive Gulliver can be lethally threatened by immense wasps and rats, but also his manhood and sexuality are equally disparaged. Gulliver accordingly finds himself humiliated when the queen's ladies-in-waiting undress, urinate, and defecate in his presence, ignoring him altogether. He simply doesn't count.

At the same time, Gulliver finds himself repulsed rather than aroused by their nakedness, because their huge size exaggerates the dimensions of their skin blemishes and gaping pores.

Although Gulliver eventually escapes from Brobdingnag, he cannot get away from his low opinion of the human species and his sense of its insignificance—a perspective shared in the present book, although for our purposes the issue is not that people are downright pernicious or odious but that when it comes to inherent significance they are literally smaller than the Lilliputians were to Gulliver, or Gulliver to the giants of Brobdingnag. It's a perspective that may well be enhanced if we ever discover signs of life on other "heavenly" bodies, or simply recognize the deeper consequences of the fact that we occupy an average planet, orbiting a rather unexciting star in an out-of-the-way corner of a comparatively trivial galaxy. And that we have come to exist as a result of purely material processes (notably natural selection combined with the laws of chemistry and physics), devoid of any deeper meaning or cosmic consequence.

This is not to say, however, that *Homo sapiens* isn't important. We are crucially and organically connected to all other life forms, which gives us a claim—although not a unique one—to a certain expansive grandeur. We are also immensely consequential for ourselves, in the same way that members of a baboon troop or a human family are important to each other. Moreover, we are important in ways that go beyond our significance as individuals and as organic beings, in that we—more, perhaps, than any other species—have already had immense practical impact on our planet and its creatures, and promise (more accurately, threaten) to do even more. Environmental scientists—beginning, it appears, with Nobel prize–winning atmospheric chemist Paul Crutzen—have argued for some time that we are living in our own human-created era, the Anthropocene, a time in which the cumulative effect of our activities dominates the machinery of Earth. Geologists resisted this concept, maintaining that establishing a new, recognized epoch requires not only a clear origination point but also something that constitutes a permanent and worldwide demarcation, equivalent, for example, to the extinction of the dinosaurs nearly 70 million years ago, which marked the end of the Cretaceous. Writing in the International Geosphere-Biosphere Programme newsletter in 2000, Crutzen and fellow atmospheric scientist Eugene Stoermer urged nonetheless that, given the key role played by human beings in Earth's planetary ecosystem, the concept of Anthropocene ("human era") was fully appropriate.

As to when precisely the Anthropocene began, Crutzen and Stoermer suggested that

to assign a more specific date to the onset of the "anthropocene" seems somewhat arbitrary, but we propose the latter part of the 18th century, although we are aware that alternative proposals can be made (some may even want to include the entire holocene). However, we choose this date because, during the past two centuries, the global effects of human activities have become clearly noticeable. This is the period when data retrieved from glacial ice cores show the beginning of a growth in the atmospheric concentrations of several "greenhouse gases", in particular CO_2 and CH_4. Such a starting date also coincides with James Watt's invention of the steam engine in 1784.[1]

Other possible markers for the commencement of the Anthropocene include the early 1950s, when atmospheric nuclear testing added layers of radioactive fallout worldwide, the (geologically speaking) nearly instantaneous accumulation of aluminum, plastic, and concrete particles—notably in the oceans, the suddenly high global soil levels of phosphate and nitrogen derived from fertilizers, and so on, even for some, the widespread appearance of domestic fowl, whose bones can now be found in geologic deposits throughout the globe. Regardless of the precise demarcation point, which has yet to be agreed, in 2016 the Working Group on the Anthropocene recommended overwhelmingly to the International Geological Congress that this new epoch be recognized.

For our purposes, the key point is that anthropodiminution is very different from "anthropodenial," a refusal to acknowledge that human beings have been exerting an immense influence—much of it malign—on the planet Earth. The Anthropocene is real. So is anthropocentrism, the conceit that figuratively, if not literally, the universe revolves around *Homo sapiens*. But anthropocentrism is "real" only in the sense that many people believe it, even though it isn't true.

Old paradigm: Human beings are fundamentally important to the cosmos.

New paradigm: We aren't.

NOTE

1. Crutzen, P., and E. Stoermer. "International Geosphere Biosphere Programme (IGBP) Newsletter, 41" (2000).

2

From Centrality to Periphery

ACCORDING TO FRANCIS BACON, in his essay *Prometheus: or the State of Man*, composed roughly five centuries ago, "Man, if we look to final causes, may be regarded as the centre of the world . . . for the whole world works together in the service of man . . . . All things seem to be going about man's business and not their own." This is classic anthropocentrism, a perspective that is comforting and not uncommon, although it is completely wrong.[1] Think of the mythical, beloved grandmother, who lined up her grandchildren and hugged every one while whispering privately to each, "*You* are my favorite!" We long to be the favorite of god or nature, as a species no less than as individuals, and so, not surprisingly, we insist on the notion of specialness. The center of our own subjective universe, we insist on being its objective center as well. This is the same error that led Thomas Jefferson to react as follows to the discovery of fossil mammoth bones in his beloved Virginia: "Such is the economy of nature, that no instance can be produced of her having permitted any one race of animals to become extinct." And maybe, even now, in some as yet undiscovered land, there are modern mastodons, joyously cavorting with giant sloths and their ilk, testimony to the unflagging concern of a deity or at minimum, a natural design, that remains devoted to all creatures . . . especially, of course, ourselves. Just don't count on it.

It might be useful to introduce a new word to go with anthropocentrism: "modocentrism" (derived from the Latin for "now" or "current"). Modocentrism would refer to the peculiarly widespread idea that modern times—or whatever era or date is currently in question—are unique in history. Of course, as Heraclitus pointed out, one cannot step in the same river twice; every situation and every slice of time is unique unto itself. However, there seems to be a widespread illusion that present days are truly exceptional, either in how good they are, or (more often) how bad, also in how interesting they are, how consequential, and so forth. Modocentrism is thus equivalent to each of the grandmother's "favorites" believing that he or she is not only special but also occupies a unique and notable time frame.

Modocentrism is operating when people announce—as they have apparently been doing throughout history—that children have never been as ____ (fill in the blank) as they are these days. Or that we are living in a time of extraordinary peril, in particular because of the danger posed by global climate change and the threat of nuclear apocalypse. In fact, a strong objective case can and should be made that because of these twin looming disasters, we are indeed living in a time of extraordinary peril! Maybe this is yet another example of modocentrism, in which case it reveals how encompassing such a perspective can be; on the other hand, just as Freud is reputed to have noted that sometimes a cigar is just a cigar, sometimes what appears to be a time of unique danger may really be a time of unique danger. Identifying the Anthropocene might in any event be seen as a case of modocentrism. But just as paranoids, too, can have enemies, sometimes modocentrism is correct.

It is easy, nonetheless, to fall victim to erroneous modocentrism, as with the assertion that the twenty-first century is unique, for example, in how disrespectful children are to their parents, how busy people are (or how much free time they have), how bombarded they are with information, how disconnected they are from each other on a human level, how much violence we experience (or how peaceful most human lives have become), compared with some—or any—other times. It is also tempting to conclude that modernity—at least, beginning with the physics revolution of the twentieth century—has been uniquely disorienting for humanity. Thus, quantum mechanics raises weird issues regarding cause and effect, just as relativistic mechanics does to the meaning of space, while both revolutions have effectively upended the seemingly unitary "arrow of time."

Also in the twentieth century came Alfred Wegener's theory of continental drift, first published in 1912 and originally scorned by the overwhelming majority of geologists, but now abundantly confirmed as part of our general understanding of plate tectonics. This was also roughly the same time that, as already mentioned, Freud was upending confidence in our own conscious, rational mind.

And yet, these were not the first disorienting episodes to confront our species. Especially notable in the nineteenth century was Darwin's identification of evolution by natural selection as the mechanism by which all living things—including human beings—have been and are still being "created."

It is no exaggeration to note that the most significant yet widely denied consequence of biological insights is the fact that *Homo sapiens*, like the rest of the natural world, were produced by a strictly material, altogether natural process, namely, evolution by natural selection. "Descended from apes?" the wife of a prominent Victorian bishop is reported to have expostulated. "Let us hope it isn't true! But if it is true, let us hope that it doesn't become widely known!" Well, it is true, and it is widely known among anyone scientifically knowledgeable, although according to a 2014 Gallup survey, an astonishing 42% of Americans[a] believe that God—not evolution—created human beings, and did so within the last 10,000 years.

[a] Including the vice president of the United States, elected in 2016.

Although much of the material in this chapter—and Part 1 generally—disputes traditional monotheistic conceptions of human importance, the primary target of this book is not religion as such, but rather, the pervasive, exaggerated conception of human centrality. This viewpoint isn't only characteristic of much religion but also may well be deeply ingrained in the consciousness of our species, analogous to what psychoanalysts designate "infantile omnipotence."

Nonetheless, as Darwin noted in *The Descent of Man and Selection in Relation to Sex* (1871):

> The great principle of evolution stands up clear and firm, when . . . facts are considered in connection with others, such as the mutual affinities of the members of the same group, their geographical distribution in past and present times, and their geological succession. It is incredible that all these facts should speak falsely. He who is not content to look, like a savage, at the phenomena of nature as disconnected, cannot any longer believe that man is the work of a separate act of creation. He will be forced to admit that all [facts of nature] . . . point in the plainest manner to the conclusion that man is the codescendant with other mammals of a common progenitor.

At about the same time that Darwin was upending humanity's sense of itself, things were at least as destabilized in the social sphere. As Marx and Engels wrote in *The Communist Manifesto*,

> All fixed, fast-frozen relations, with their train of ancient and venerable prejudices and opinions, are swept away, all new-formed ones become antiquated before they can ossify. All that is solid melts into air, all that is holy is profaned, and man is at last compelled to face with sober senses his real conditions of life, and his relations with his kind.

Written in 1848, just as violent revolutions (all of them ultimately suppressed) were convulsing Europe, the *Manifesto* was both a cause of and a response to those shaky times that characterized the middle of the nineteenth century, an era that from the perspective of twenty-first-century modocentrism may well seem comparatively calm and quiet, even boring.

Nineteenth-century Marxists weren't the only people who felt discombobulated, that solidity was melting into air. Before the nineteenth century came what is widely known as the Enlightenment, when many prior ideas—including but not limited to the legitimacy of religion itself—were subjected to the hard light of reason. As Alexander Pope saw it, in his "Epigram on Sir Isaac Newton," "Nature's Laws lay hid in Night: God said,

'Let Newton be!' and all was light." Alas, not quite. Lots of things remained hidden post-Newton, just as the Enlightenment itself had been preceded by some truly profound jostling, notably replacement of the geocentric worldview of Ptolemy with the heliocentric perspective identified by Copernicus, Kepler, and Galileo. It is difficult, perhaps impossible, for inhabitants of the twenty-first century to appreciate the profound sense of disorientation that resulted, and that led many a well-informed person of his or her time to despair that things had never been so confusing, that the place of *Homo sapiens* had never been so unmoored. Modocentrism indeed.

The following, from John Donne's 1611 poem, "The Anatomy of the World," expresses the sense of loss verging on betrayal occasioned by advances in astronomy at the time:

> The sun is lost, and th'earth, and no man's wit
> Can well direct him where to look for it.

Eventually, even as we—as a species—found the Earth once again, and came to accept its uninspiring position as the third planet out of nine[b] going around a sun, which itself isn't especially notable, in a decidedly noncentral location within a mediocre galaxy (the Milky Way), humanity's sense of itself began to reel once again, not so much from the insights of astronomy as from biology. For some, the loss of planet Earth's centrality remained a potent metaphor of disorientation. "What were we doing," asked Nietzsche in *The Gay Science*, "when we unchained this earth from its sun? . . . Are we not plunging continually? Backward, sideward, forward, in all directions? Is there still any up or down?"[2]

Tempting as it might be to console such anguish with assurance that there still is an up and a down, reality is otherwise. Of course, up and down persist in everyone's immediate environment, but it is more than trivially true that the direction "down" at any point on Earth, if continued through the planet to the opposite side, becomes "up." In fact, the standard picture of our globe—with Canada and the United States up and Latin America down, Europe up, and Africa down—is simply a North-centric self-congratulation. It would be every bit as accurate, geographically (even as it is deflating, ethnocentrically for those of us in the Northern Hemisphere) to reverse this perspective, and make the Southern Hemisphere "up" and the Northern "down."

Nietzsche's sense of "plunging continually" is even greater if you move into the solar system and more yet if you enter deep space, where there absolutely is no up or down. Although this insight is dizzying for some, in a sense it is less disruptive

[b] Now eight, with the demotion of Pluto.

than the one Nietzsche was responding to, and that underpins—and for many people, undermines—everything we know about ourselves.

People have long speculated about whether we Earthlings are alone in the universe. The Inquisition burned Giordano Bruno at the stake in part because his speculation about the existence of extraterrestrial life didn't accord with the Catholic Church's sixteenth-century assessment. According to the astronomer Percival Lowell, extraterrestrial life was not only possible but to be expected: "We know life to be as inevitable a phase of planetary evolution as is quartz or feldspar or nitrogenous soil. Each and all of them are only manifestations of chemical affinity."[3] Although Lowell erred mightily in proclaiming that Mars had canals and thus, intelligent inhabitants who engineered them, his perception of life as a result of "chemical affinity" has much to recommend it.

If genuine ETs (extraterrestrials) are ever found, would this further threaten the belief that the Earth is a privileged place and that we are special and sacred? It should. It would also pose some intriguing problems for Earth-bound religions, nearly all of which are literally Earth- and human-bound. Might Christians conclude that Christ and the Christian god are limited to Earth, perhaps with other deities relevant to extraterrestrials? Or would Jesus and God—who according to most interpretations of scripture are especially associated with our planet—hold dominion elsewhere as well? This would seem likely, if only because the world's monotheistic religions tend to be overwhelmingly totalistic, presuming that the one all-powerful god, insofar as he is truly all-powerful, reigns everywhere in the universe, even though other planets aren't mentioned in the Bible. If so, then Earth would remain, in a deep theological sense, the literal center of the cosmos even as its astronomical significance has long ago been demoted. This would likely be news to any extraterrestrials, who could well see things differently.

Let's pursue this a bit farther. Would ET possess original sin, or is this limited to the descendants of Adam and Eve? Certain faiths would have less trouble with this question; others more. The Seventh Day Adventist prophet Ellen White claimed to perceive extraterrestrials who were "tall, majestic people," sin-free, and not in need of redemption. Jews will almost certainly not feel moved to convert ET, since they aren't even particularly eager to convert regular terrestrials. By contrast, Roman Catholics, Muslims, and Mormons presumably will be eager to spread their religion beyond Earthlings, given that they enthusiastically—even urgently—seek converts among this planet's inhabitants. The Qur'an states that "all of the universe's beings serve Allah, who takes account of them all, and has numbered them exactly," so if nothing else the mere presence of extraterrestrials might not occasion great surprise and consternation among the followers of Mohammed.[4]

Buddhists are not especially prone to proselytizing, but they would probably have no problem with the existence of extraterrestrials. Buddhist cosmology maintains that the universe is unimaginably huge and ancient, and that souls transmigrate at death to various beings in all possible places in the universe. So, human souls might well occupy ET bodies and vice versa: we might even be temporary abodes for ET souls. It should go without saying that an atheistic evolutionary perspective will have no trouble whatever with alien extraterrestrial life, since there is every reason to think that the basic principles of natural selection would operate on any planet and in any environment, and no reason to expect that the Earth, and the creatures thereon, are either especially privileged or exempt.

Old paradigm: We are literally central to the universe, not only astronomically but in other ways, too.

New paradigm: We occupy a very small and peripheral place in a not terribly consequential galaxy, tucked away in just one insignificant corner of an unimaginably large universe.

NOTES

1. Some material in this chapter has been altered and repurposed from my article "Why We Aren't So Special," appearing in *The Chronicle of Higher Education* on January 3, 2003.
2. Friedrich Nietzsche, *The Gay Science* (Cambridge: Cambridge University Press, 2001).
3. Percival Lowell, *Mars as the Abode of Life* (London: Macmillan, 1909).
4. David Weintraub, *Religions and Extraterrestrial Life* (New York: Springer, 2014).

3

The Meaning of Life

AT ONE POINT in Douglas Adams's hilarious *Hitchhiker's Guide to the Galaxy*, a baby sperm whale has some thoughts as it plummets toward the planet Magrathea.[1] This appealing but doomed animal had just been "called into existence" several miles above the planet's surface, when one of two nuclear missiles, directed at our heroes' space ship, had been inexplicably—and indeed, improbably—transformed via an "Infinite Improbability Drive." (The other missile was turned into a bowl of petunias.) I'll let the masterful and much-missed Mr. Adams take it from here:

> And since this is not a naturally tenable position for a whale, this poor innocent creature had very little time to come to terms with its identity as a whale before it then had to come to terms with not being a whale any more. This is a complete record of its thoughts from the moment it began its life till the moment it ended it. "Ah . . . ! What's happening?" it thought. "Err, excuse me, who am I?" "Hello?" "Why am I here? What's my purpose in life?" . . . "Never mind, hey, this is really exciting, so much to find out about, so much to look forward to, I'm quite dizzy with anticipation" . . . "And wow! Hey! What's this thing suddenly coming towards me very fast? Very fast. So, big and flat and round, it needs a big wide sounding name like . . . own . . . found . . . round . . . ground! That's it! That's a good name—ground!" "I wonder if it will be friends with me?" And the rest, after a sudden wet thud, was silence. Curiously enough, the only thing that went through the mind of the bowl of petunias as it fell was "Oh no, not again." Many people have speculated that if we knew exactly why the bowl of petunias had thought that we would know a lot more about the nature of the universe than we do now.[2]

One thing we know about the nature of the universe is that evolution, too, is an improbability generator, although its outcomes are considerably more diverse than a single sperm whale (however doomed and adorable), not to mention that bowl of petunias. But the key point for our purposes is that after we are called into existence by that particular improbability generator called natural selection, human beings have no more purpose in life than Douglas Adams's naive and ill-fated whale, whose blubber was soon to bespatter the Magrathean landscape.

Let's start with some pure biology: nobody gets out of here alive. And at the other end, nobody got here except through a chance encounter between a particular sperm (minus the whale) and a particular egg. Had it been a different sperm, or a different egg, the result would have been a different individual. Biology again. We, like all other sexually reproducing creatures, result from the conjunction of certain types of matter known as sperm and egg, nucleotides, proteins, carbohydrates, and a very large number of other purely physical entities, with nothing approaching "purpose" anywhere to be seen. Finally, as to *why* we are here, the life sciences once again have an answer: human beings, like all other beings, aren't here for any purpose that in any way transcends what their genes were up to in the first place. Evolution is a genetic process, and *all* bodies have been "created," unlike Adams's Magrathean whale, for no purpose except the dissemination of certain genes.

Admittedly, there isn't much in gene propagation itself to make the heart sing. And in an increasingly overcrowded, polluted, and resource-depleted world there is much reason to deny its prodding. It is not something to sneer at, however; after all, every one of your direct ancestors has reproduced, without ever missing a beat, going back to the primordial ooze. But at the same time, no one likes to be manipulated, even when the manipulator is our own DNA! As Richard Dawkins emphasized so dramatically at the end of *The Selfish Gene*, it is also well within the human repertoire to rebel against our evolutionary purpose(lessness), thereby saying "No" to our genes.

Homo sapiens is probably the only life form with this capability, and indeed, the human search for meaning has been as persistent as it is inchoate. Where, then, does biological insight leave the human search for meaning? I see two fundamental possibilities. On the one hand, we can delude ourselves, clinging to the infantile illusion that some One, some Thing, is looking over us, somehow orchestrating the universe with each of us personally in mind. Or, we can face, squarely, the reality that life in general and our individual life in particular is inherently meaningless.

Here is a forthright acknowledgment from Heinrich Heine. In his poem "Questions," we are introduced to someone who asked the waves, "What is the meaning of Man? Whence did he come? Whither does he go? Who dwells up there on the golden stairs?" And in response: "The waves murmur their eternal murmur, the wind blows, the clouds fly, the stars twinkle, indifferent and cold, and a fool waits for an answer."

This does not imply giving up on the search for meaning. Quite the contrary, it italicizes the foolishness of waiting for the world to provide an answer, expecting it (the waves, the wind, the clouds, the stars, our fellow creatures, an imagined deity, or a human-crafted written text) to reveal our meaning or purpose, as though these somehow exist outside ourselves, just waiting to be uncovered. Instead of despair, this perspective opens a creative locus of compatibility: between a recognition of life's biologically based meaninglessness and another recognition, of the responsibility for people to achieve meaning in their lives—not by hiding behind the dictates of dogma, or the promise of a "purpose" preprogrammed for each individual, but by how each of us chooses to live his or her life in a world that is inherently lacking in purpose.

Call it a kind of evolutionary existentialism. In an absurd, inherently meaningless world—our unavoidable evolutionary legacy as material creatures in a physically bounded universe—the only route to meaning is to achieve it by how we engage our own sentient existence. This vision of life's absurdity is not surprising. It is, in fact, altogether appropriate, given that human beings—just as all other living things—are the products of a mindless evolutionary process whereby genes joust endlessly with other genes to get ahead. "Winners" are simply those who happen to be among those left standing whenever a census is taken, but how shallow that the only "goal" is to stay in the game as long as possible! Moreover, it is ultimately a fool's game, in which we and our DNA can never cash in our chips and go home.

Death, as the existentialists insist on pointing out, makes life absurd. The only thing more absurd is to deny the absurdity, to be stuck in a meaningless life, recognizing the absence of meaning only dimly, if at all, pretending that Mommy or Daddy, Jehovah, Allah, or Brahma has everything planned out—just for us.

In his celebrated and influential book *Natural Theology* (1803), William Paley wrote as follows about cosmic beneficence and species centrality:

> The hinges in the wings of an earwig, and the joints of its antennae, are as highly wrought, as if the Creator had had nothing else to finish. We see no signs of diminution of care by multiplication of objects, or of distraction of thought by variety. We have no reason to fear, therefore, our being forgotten, or overlooked, or neglected.

Earlier, we noted Thomas Jefferson's comparable, well-intended, but totally inaccurate reassurance that no animals have ever gone extinct, and accordingly, we *Homo sapiens* shouldn't lose heart. Just as there are thirty different species of lice that make their homes in the feathers of a single species of Amazonian parrot, each of them doubtless put there with *Homo sapiens* in mind, we can be confident that our existence is so important that we would never be ignored or abandoned.[a] An accomplished amateur paleontologist, Jefferson remained convinced that there must be mammoths

[a] Human beings host three different species: head lice, body lice, and pubic lice. Whether this, too, should be seen as an indication of divine benevolence (i.e., that we have only three), is a personal judgment.

lumbering about somewhere in the unexplored arctic regions; similarly with the giant ground sloths whose bones had been discovered in Virginia, and which caused consternation to his contemporaries.

In his *Physics* and *Metaphysics*, Aristotle famously distinguished between the different kinds of causation, notably "final" versus "efficient" causes, the former being the goal or purpose of something, and the latter, the immediate mechanism responsible. The evolutionary biologist Douglas Futuyma[3] has accordingly referred to the "sufficiency of efficient causes." In other words, since Darwin, it is no longer useful to ask, "Why has a particular species been created?" It is not scientifically productive to assume that the huge panoply of millions of species—including every obscure soil microorganism and each parasite in every deep-sea fish—exists with regard to and somehow because of human beings. Similarly, it is no longer useful to suppose that we, as individuals, are the center of the universe, either. Efficient causes, the material factors that generate a particular outcome (aka, the working of causes and effects) are enough.

"We find no vestige of a beginning," wrote the pioneering geologist James Hutton, in his *Theory of the Earth* (1788), "no prospect of an end." For some, this prospect is bracing; for others bleak, even terrifying. The seventeenth-century mathematician and Catholic mystic Blaise Pascal, gazing similarly into a vastness devoid of human meaning or purpose, wrote in Pensées,

> When I consider the short duration of my life, swallowed up in an eternity before and after, the little space I fill engulfed in the infinite immensity of spaces whereof I know nothing, and which know nothing of me, I am terrified. The eternal silence of these infinite spaces frightens me.

Of course, maybe I am wrong, and Hutton too, and also Darwin, as well as Copernicus. Perhaps Pascal should have been encouraged rather than frightened. And maybe the whale of Magrathea was onto something. Perhaps each of us is genuinely central to some cosmic design. Many people contend that they have a personal relationship with god; for all I know, maybe god reciprocates, tailoring his grace to every such individual, orchestrating each falling sparrow and granting to every human being precisely the degree of importance that so many crave. Maybe we have a role to play, and maybe—as so many people in distress like to assure themselves—they will never be given more than they are capable of bearing. Maybe we aren't Magrathean whales after all, flopping meaninglessly and doomed to fall. The world's cemeteries, however, suggest otherwise.

THERE IS SOME reason to think that the stubborn persistence of religion is due at least in part to our equally stubborn search for significance, given how small and seemingly trivial each of us appears in the context of the universe, and even among the numerous life forms on our own planet. If so—if a large part of religion's appeal lies in its promise of special significance for *Homo sapiens* in general and for each of us in particular— then here is some bad news, for religion as well as for the many individuals motivated by this quest: we simply aren't especially significant. It's that simple. Or at least, no one is significant just because he or she exists. If we want significance in our lives, we have to earn it. Meaning isn't bestowed upon anyone merely because there is a god who gives a damn, or by virtue of existence itself. Lots of living things exist; mere existence is a dime a dozen. Or a billion.

In the right hands, there can be genuine humor in life's meaninglessness. Of the now vast "literature of the absurd"—much of it theater—a large percentage is downright funny: black humor, to be sure, but humor nonetheless, including, most notably, that of Samuel Beckett. As Beckett's novel *Murphy* draws to its antic conclusion, we are given a memorable account of what ultimately became of the hero's ashes, and thus, of human goals and aspirations more generally (it is said that Beckett's mentor, James Joyce, was so fond of this passage that he committed it to memory):

> Some hours later Cooper took the packet of ash from his pocket, where earlier in the evening he had put it for greater security, and threw it angrily at a man who had given him great offence. It bounced, burst, off the wall on to the floor, where at once it became the object of much dribbling, passing, trapping, shoot-ing, punching, heading and even some recognition from the gentleman's code. By closing time the body, mind and soul of Murphy were freely distributed over the floor of the saloon; and before another dayspring greened the earth had been swept away with the sand, the beer, the butts, the glass, the matches, the spits, the vomit.[4]

Beckett's best-known work, *Waiting for Godot*, is constructed of similarly shocking and occasionally brilliant absurdities. Like *Murphy*—and like those willing to adopt an evolutionary perspective in its entirety—it confronts the reality of meaninglessness. Two tramps, Vladimir and Estragon, spend all of Act I waiting for Godot, with whom they believe they have an appointment. He doesn't show up. Nor does he appear in Act II, leading to one wag's description of the play as one in which nothing happens. Twice.

Toward the end, Vladimir cries to his buddy, Estragon: "We have kept our appoint-ment, and that's an end to that. We are not saints, but we have kept our appointment. How many people can boast as much?" Vladimir replies lugubriously: "Billions." After all, there is nothing unusual in keeping one's appointment. We all do it, insofar as we exist. Our friend the whale similarly showed up for his appointment with the ground. (Woody Allen once noted that 99% of life is just showing up; biologically, it is 100%.)

Maybe the key, then, is what you do while you are waiting. This is precisely what underlies the literature of existentialism and what I am calling evolutionary existentialism: the prospect of redemption, of achieving meaning via meaningful behavior, even though—or rather, especially *because*—in the long run any action is meaning*less*.

One of the greatest such accounts, and one of the best examples in any novel of people achieving meaning through their deeds, is Albert Camus's *The Plague*, which describes events in the Algerian city of Oran during a typhoid epidemic. In this novel, the plague stands for many things, among them the German occupation of France during World War II, the inevitability of death, and, of course, plague itself. The ultimately futile struggle against cruelty, against death and disease, against an uncaring universe, is testimony to the fact that our creaturely existence is not, in itself, meaningful. After all, not only is it true that "no one gets out of here alive," but cruelty and disease (along with death) have a habit of reappearing.

What gives life meaning, therefore, is not the prospect of ultimate success but *how* one chooses to live. *The Plague* is a "chronicle" compiled by the heroic Dr. Rieux, in order to "bear witness in favor of those plague-stricken people; so that some memorial of the injustice and outrage done them might endure; and to state quite simply what we learn in a time of pestilence: that there are more things to admire in men than to despise."[5]

But Rieux, and Camus himself, is aware that we live always in a time of pestilence. Camus's most famous essay, "The Myth of Sisyphus," makes the point that Sisyphus stands for all humanity, ceaselessly pushing our rock up a steep hill, only to have it roll back down again. That is our lot. And over and over, generation after generation, death after life after death after life, that is the lot of our genes. Camus concludes his essay with the stunning announcement that "One must imagine Sisyphus happy," because he accepts his rock and his hill, defining himself—achieving his own meaning—within its constraints. Upon reflection, it should be apparent that Camus's stance, in which meaning is not conveyed by life itself but must be imposed on it, is not only consistent with an informed biological perspective, it is actually the only thing—in light of that perspective—that makes sense.

For Camus, the "absurdity" of life derives from the confrontation between the all-too-human longing for meaning and the "unreasonable silence of the world." Despite the fact that his labor will never succeed, Sisyphus isn't simply absurd; rather, he is an "absurd hero," whose heroism derives specifically from his awareness that his project is an impossible one, that the rock will never remain perched triumphantly atop his mountain. If he were thus deluded, he would simply be absurd—indeed, ridiculous—rather than heroic. It is precisely because he knows that his labor is inherently meaningless, but he does it anyhow, that Sisyphus becomes heroic and, paradoxically, achieves meaning. (Another way of looking at Sisyphus, proposed in jest by one of my students, is that perhaps he had suffered a stroke and experienced a persistent failure of short-term memory, regularly forgetting that he "failed" last time!)

But as Camus puts it, Sisyphus knows full well what he is doing, and he persists in his project with no illusions and thus, with uniquely human heroism:

> Sisyphus teaches the higher fidelity that negates the gods and raises rocks. He too concludes that all is well. This universe henceforth without a master seems to him neither sterile nor futile. Each atom of that stone, each mineral flake of that night filled mountain, in itself forms a world. The struggle itself toward the heights is enough to fill a man's heart.

As the existentialists have long recognized, the quest for meaning itself only takes on meaning—indeed, it only takes place at all—because people are creatures whose lives, embedded in biology, are fundamentally lacking in meaning. As Darwin noted toward the end of *The Origin of Species*, "There is grandeur in this view of life." Moreover, if life's purpose were evident, existence would be devoid of some of its greatest triumphs, which show human beings struggling to make sense of a purposeless world: not how people rise above their biology (for that cannot be done), but how they ride their biology to greater heights, or, naively hopeful, like the whale of Magrathea, simply plummet unknowingly to their deaths.

At the conclusion of *The Plague*, while the citizens of Oran are celebrating their "deliverance" from the epidemic, Dr. Rieux knows better. But at the same time, his commitment to the struggle, to what defines human beings in an otherwise uncaring universe, is undiminished. Rieux understands that in the game of life, all victories are temporary, which renders his perseverance all the grander:

> He knew that the tale he had to tell could not be one of a final victory. It could be only the record or what had had to be done, and what assuredly would have to be done again in the never-ending fight against terror and its relentless onslaughts, despite their personal afflictions, by all who, while unable to be saints but refusing to bow down to pestilence, strive their utmost to be healers.[6]

For many, connecting evolutionary biology and existentialism seems oxymoronic in the extreme. I maintain, by contrast, that they are natural allies. In italicizing our lack of inherent meaning while striving to respond to our very human, biologically generated need for it, practitioners are, in their own way, refusing to bow down to some of evolution's more unpleasant imperatives. Although no one can announce final victory, Sisyphus might well applaud such efforts.

Old paradigm: Each human life has its own inherent meaning; it is up to every person to discover it.

New paradigm: No one's life is automatically endowed with any meaning, simply by virtue of his or her existence; it is up to individuals who seek meaning to define and establish it by how they live.

NOTES

1. Some material in this chapter has been modified from "What the Whale Wondered: Evolution, Existentialism, and the Search for Meaning," my contribution to *Richard Dawkins: How a Scientist Changed the Way We Think*, ed. A. Grafen and M. Ridley (London: Oxford University Press, 2006).
2. Douglas Adams, *The Hitchhiker's Guide to the Galaxy* (London: Pan, 1985).
3. Douglas Futuyma, *Science on Trial: The Case for Evolution* (Sunderland, MA: Sinauer Associates, 1995).
4. Samuel Beckett, *Murphy* (New York: Evergreen Press, 1957).
5. Albert Camus, *The Plague*, trans. Stuart Gilbert (New York: The Modern Library, 1962).
6. Camus, *The Plague*.

4

Well Designed?

EVEN THOUGH WE aren't the literal (or even metaphoric) center of the universe, even though we weren't created specially as chips off the Old Divine Block, and even though our lives haven't been endowed with personalized, prefabricated meaning, at least we are well put together, testimony to evolution's remarkable powers. Or so it might seem.[1]

In 1829, Francis Henry Egerton, the 8th Earl of Bridgewater, bequeathed eight thousand pounds sterling to the Royal Society of London to support the publication of works "On the Power, Wisdom, and Goodness of God, as Manifested in the Creation." The resulting *Bridgewater Treatises*, published between 1833 and 1840, were classic statements of "natural theology," seeking to demonstrate God's existence by examining the "perfection" of the organic world.

These days, biologists are often inclined to keep pointing, albeit with different motivation, to the extraordinary complexity and near perfection of living things as evidence of the extraordinary effectiveness of natural selection, as manifested in evolution. Such gestures are understandable, even laudable, contributing as they do to a healthy "gee-whiz" appreciation of the Darwinian process and the subtle, complex, and highly effective organic world that it has produced. But ironically, they are less useful than one might think, especially in distinguishing natural selection from its premier alternative (at least among the biologically illiterate): special creation, or, in its barely disguised incarnation, "intelligent design theory."

The problem is that those same wonders of perfection used by biologists to buttress others' confidence in natural selection can also be used by believers in so-called intelligent design as evidence for a divine designer. Fortunately, however, the two explanations are in fact discriminable, and not simply because natural selection provides a predictable and testable mechanism whereas resort to miracles explains everything and thus, nothing. Some of the most powerful distinctions between a theological and a biological perspective are provided not by the perfection of living things, but by their *imperfection*. Thus, it is worth emphasizing that even though natural selection regularly produces marvels of improbability (a living thing is, above all else, tremendously nonrandom and low-entropy), it is necessarily a blundering, imperfect, and highly unintelligent engineer, as compared to any purportedly omniscient and omnipotent creator. Ironically, it is the stupidity and inefficiency of evolution—its manifold design-flaws—that argue most strongly for its material and wholly earth-bound nature.

Theodosius Dobzhansky, one of the most influential geneticists of the twentieth century, wrote an article with an oft-quoted title: "Nothing in Biology Makes Sense Except in the Light of Evolution."[2] Let's reformulate this more positively: "Lots of things in biology—including but not limited to human biology—make perfect sense in the light of evolution." This is especially true, paradoxically, of our imperfections.

Natural selection is a mathematically precise process, whose outcome should be—and for the most part, is—a remarkable array of "optimal" structures and systems. A naive view therefore assumes that the biological world is essentially perfect and certainly highly predictable, like a carefully orchestrated geometric proof. Or a billiard game, in which a skilled player can be anticipated to employ the correct angles, inertia, force, and momentum. True to expectation, living things reveal some pretty fancy shooting. Specialists, no less than biologically informed laypeople, are therefore inclined to applaud the outcome, and rightly so.

In his *Dialogues Concerning Natural Religion* (1791), even David Hume—arch skeptic, materialist, and atheist—marveled at how the parts of living things "are adjusted to each other with an accuracy which ravishes into admiration all men who have ever contemplated them."[3]

Such admiration, however, is not always warranted. Gilbert and Sullivan's "Mikado" sings about letting "the punishment fit the crime," gleefully announcing, for example, that the billiard sharp will be condemned to play "on a cloth untrue, with a twisted cue, and elliptical billiard balls." To a degree not generally appreciated, the organic world contains all sorts of imperfections, and as a result, shots often go awry—not because the laws of physics and geometry aren't valid, or because the player isn't skillful, but because even Minnesota Fats was subject to the sting of reality.

Make no mistake, evolution—and thus, nature—IS wonderful. The smooth-running complexity of physiological systems, anatomical structures, ecological interactions, and behavioral adjustments are powerful testimony to the effectiveness of natural selection in generating highly nonrandom systems such as the near-incredible complexity of the human brain, the remarkable lock-and-key fit between organism and environment, antibody and antigen, and the myriad interconnecting details of how a cell reproduces itself, extracts energy from complex molecules, and so forth.

This does not, however, support the Panglossian view that "all is for the best in this best of all possible worlds," nor does it offer aid and comfort to those seeking evidence of divine benevolence in the machinery of the natural world. In a much-quoted letter to his supporter, the American botanist Asa Gray, Darwin wrote,

I own that I cannot see, as plainly as others do, & as I should wish to do, evidence of design & beneficence on all sides of us. There seems to me too much misery in the world. I cannot persuade myself that a beneficent & omnipotent God would have designedly created the Ichneumonidæ [parasitic wasps] with the express intention of their feeding within the living bodies of caterpillars, or that a cat should play with mice.

DARWIN WAS RATHER tenderhearted. Evolution is not.

Moreover, not only is there no evidence for an underlying benevolence in evolution, or anything approaching human ethics or morality, but also, on a strictly physical level of practical design, there are abundant imperfections. For now, let's concentrate on just one dimension, and moreover, on just one species: *Homo sapiens.*

Among evolution's numerous constraints, one of the most vexing, and unavoidable, is history, the simple fact that living things have not been created *de novo*, but rather, have evolved from antecedents. If they were specially and intelligently designed in each case, there is no reason for the designer not to have chosen the optimum pattern every time. On the other hand, insofar as they are constrained by their past, and are therefore the products of small incremental steps, each lacking in foresight, living things are necessarily jerry-built and more than a little ramshackle. (It might, for example, be optimal if elephants could fly. After all, because of local overpopulation in increasingly threatened game parks, many elephants are undernourished, even starving, but for some reason they are unable to hover 30 feet above the ground and eat leaves currently beyond their reach. Walt Disney's *Dumbo* notwithstanding, the evolutionary past of today's pachyderms severely constrains their present and future.)

But I promised some human examples. Here goes.

Consider the skeleton. Ask yourself, if you were designing the optimum exit for a fetus, would you engineer a crazy route that passes through the narrow confines of the pelvic girdle? Add to this the tragic reality that childbirth is not only painful in our species but also downright dangerous and sometimes lethal, owing to occasional cephalopelvic disproportion—literally, the baby's head being too large for the mother's birth canal—along with breech presentation, and so forth. This design flaw is all the more dramatic since there is plenty of room for even the most stubbornly misoriented (i.e., breech) or large-brained fetus to be easily delivered, any place in that vast nonbony region of a woman's body below the ribs and above the pelvis. And in fact, that is precisely what obstetricians do, when performing a Cesarean section.

Evolution, however, stubbornly and stupidly insisted on threading its way through the ridiculously narrow pelvic ring, altogether neglecting the simple, straight-forward solution, which would have been for the vagina to open pretty much anywhere else in the lower abdomen. Why? Because evolution isn't an observing, creating, all-knowing engineer and intelligent designer. Rather, it is a mechanical, mathematically consistent but completely unconscious and natural process. Among its constraints is the fact that species aren't "created" out of whole cloth; rather, they evolve—slowly and imperfectly—from their ancestors.

Human beings are mammals, and therefore tetrapods by history. As such, our ancestors carried their spines parallel to the ground; it was only with our adaptive insistence on upright posture[a] that the pelvic girdle had to be rotated, thereby making a tight birth-fit out of what for other mammals is nearly always an easy passage. An engineer who designed such a system from scratch would get a failing grade, but evolution didn't have the luxury of design, intelligent or otherwise. It had to make do with the materials available. (Admittedly, it can be argued that the dangers and discomforts of childbirth were preplanned after all, since Genesis gives us God's judgment on Eve, that as punishment for her disobedience in Eden, "in pain you shall bring forth children." Might this imply that if Eve had only restrained herself, her vagina would have been where every woman's belly button currently resides?)

On to men. An especially awkward design flaw of the human body—male and female alike—results from the close anatomical association of the excretory and reproductive systems, a proximity attributable to a long-standing, primitive vertebrate connection, and one that isn't only troubling for those who are hygienically fastidious about their sex lives. In addition, although there is no obvious downside to the deplorable fact that the male urethra does double-duty, carrying both semen and urine, most elderly men have occasion to regret that the prostate gland is closely applied to the bladder, so that enlargement of the former impinges awkwardly on the latter. In addition, as human testicles descended—both in evolution and in embryology—from their position inside the body cavity, the vas deferens, which connects testis to urethra, became looped around the ureter (which carries urine from kidneys to bladder), resulting in an altogether ridiculous arrangement that would never have occurred if evolution could have anticipated the problem and, like an even minimally competent structural engineer, designed male tubing to run in a direct line.

In this regard, the most dramatic example of a ridiculous, deeply unintelligent, and unplanned anatomical detour orchestrated by evolution occurs in the neck of giraffes. Probably the most famous thing about the giraffe's neck (at least among biologists), is a peculiarity of its innervation, notably its left recurrent laryngeal nerve, which turns out to be a stunning example of extraordinarily dumb design . . . once again, precisely what we'd expect in a creature that, like all other creatures, wasn't "designed" at all, but is a ramshackle product of selection acting on the biological material that was historically available.

---

[a] Interestingly, although there are numerous hypotheses as to why our ancestors evolved bipedalism—that is, the adaptive payoff of being upright beings—that issue is currently unresolved.

Here's the deal: the laryngeal nerves, present in vertebrates generally, branch off from the larger vagus nerve and connect the brain to the muscles of the larynx. (Forget, for the moment, that giraffes are probably the quietest of any large mammal; they do vocalize a little, albeit faintly.) In all mammals, the recurrent laryngeal nerves depart from the vagus at the level of the aortic arch, the spot where the aorta, initially ascending from the heart and continuing via the carotid arteries to nourish the head and neck, dives posteriorly to provide blood flow to the rest of the body. This arch of the aorta makes a hairpin, 180-degree loop, which is no problem for the right recurrent laryngeal nerve, which, being on the "correct" side, goes directly up to the larynx, along the trachea. But its left counterpart is forced to curve under the aortic arch before heading larynx-ward—a bit anatomically inconvenient but not a major problem in most vertebrates including human beings, since this literally loopy path only necessitates a few extra inches of length. But herein lies both an interesting dilemma for long-necked creatures as well as an object lesson in evolution's often erroneous "design."

Among fish, the recurrent laryngeal nerves (left and right) follow a straight path from the brain, along the heart and then to the gills; pretty much the same, we can predict with near certainty, in short-necked prehistoric mammals, although the left version, stuck on the downward curving side of the aortic arch, would have had a slightly longer, loopier route. But among those critters that evolved long necks—all the better to get leaves high on acacia trees, my dear—with the heart essentially sinking low into the thorax and the larynx staying relatively high in the throat, the poor left recurrent laryngeal nerve was forced to perform a downright ridiculous detour during embryonic development: emerging from the brain, going southward so as to loop just below the ever-retreating aortic arch, then literally heading upward again, along the trachea to reach the larynx. In the case of modern giraffes, this absurd arrangement now necessitates a nerve that's about 15 feet long (7.5 feet down and then back up), whereas if it had simply been routed directly, its entire length would have been perhaps six inches! And why? Because just like our own evolution, that of giraffes wasn't laid out on a blank drawing-board; rather, it proceeded from their immediate antecedents, whose evolution derived in turn from theirs, going back to a shared ancestral fish, whose left recurrent laryngeal nerves were perfectly reasonable, thank you. (And by the way, don't expend all your recurrent laryngeal sympathy on giraffes: there were other vertebrate descendants of fish—notably the sauropod dinosaurs—whose 45-foot-long necks would have necessitated a lot more nerve: roughly 90 feet.)

Back to our own species, for a final example, although many more are available. The primitive vertebrate system, still found among some of today's chordates, combined both feeding and respiration, just as excretion and reproduction used to overlap, and still do in many species. Water went in, food was filtered out, and passive diffusion sufficed for respiration. As body size increased, a separate respiratory system was added, not *de novo* but by piggybacking onto the preexisting digestive plumbing. By consequence, access to what became the lungs was achieved only by sharing a common anteroom with incoming food. As a result, people are vulnerable to choking. The Heimlich maneuver is a useful innovation, but it wouldn't be needed if evolution simply had the foresight to design separate passages for food and air, instead of combining the two. But here as in other respects, natural selection operated by small, mindless increments, without the slightest attention to any bigger picture or anything approaching a wise, benevolent overview. It still works that way.

It must be emphasized that the preceding does not constitute an argument against evolution; in fact, quite the opposite. Thus, if living things (including human beings) were the products of special creation rather than of natural selection, then the flawed nature of biological systems, including ourselves, would pose some awkward questions, to say the least. Granted, God isn't typically conceived as giraffe-like. But if God created "man" in his image, does this imply that He, too, has comparably ill-constructed knee joints, a poorly engineered lower back, a dangerously narrow birth canal, and ridiculously ill-conceived urogenital plumbing? A novice engineer could have done better. The point is that these and other structural flaws aren't "antievolutionary" arguments at all, but rather cogent statements of the contingent, unplanned, entirely natural nature of natural selection. Evolution has had to make do with an array of constraints, including—but not limited to—those of past history.

We are profoundly imperfect, neither more nor less than all other creatures, and in these imperfections reside some of the best arguments for our equally profound naturalness.

Old paradigm: The human body is a wonderfully well constructed thing, testimony to the wisdom of an intelligent designer.

New paradigm: Although there is much in our anatomy and physiology to admire, we are in fact jerry-rigged and imperfect, testimony to the limitations of a process that is nothing but natural and that in no way reflects supernatural wisdom or benevolence.

NOTES

1. Some material in this chapter has been modified from my book *Natural Selections: Selfish Altruists, Honest Liars and Other Realities of Evolution* (New York: Bellevue Literary Press, 2007), and from my piece in *Nautilus*, "*How Necking Shaped the Giraffe*," 2015, http://nautil.us/issue/24/error/how-necking-shaped-the-giraffe.
2. Theodosius Dobzhansky. "Nothing in Biology Makes Sense Except in the Light of Evolution," *American Biology Teacher* 35 (1973): 125–129.
3. David Hume, *Dialogues Concerning Natural Religion* (1791)

5

The Anthropic Principle

THERE IS DEBATE whether the "anthropic principle" is a scientific or a philosophical concept—or primarily a religious one. Either way, it might be the most challenging argument for human specialness. The anthropic principle is based on the suggestion that if any of an array of precise physical constants, such as the gravitational constant, the exact electric charge on the proton, the mass of electrons and neutrons, and a number of other characteristics of the universe were any different, human life would be impossible.

There are many reasons, in any event, to doubt that the universe has been fine-tuned for our benefit. For one, if such tuning has happened, what's the basis for assuming that it happened with us "in mind"? (Never mind the question "Whose mind?") It is worth noting that these various physical constants are not necessarily evidence that the universe is fine-tuned to produce human beings; it could have been generated to produce the hairy nosed wombats of Australia, or maybe those bacteria and viruses that outnumber human beings by many orders of magnitude. If so, then the impact on *Homo sapiens* was merely an unanticipated side effect.

Having earlier introduced Douglas Adams's *Hitchhiker's Guide to the Galaxy* via the whale of Magrathea, I feel it necessary to point out that in Adams's antic novel, mice are "hyper-intelligent pan-dimensional beings" who are themselves responsible for the creation of the Earth. Moreover, in *The Salmon of Doubt*, Adams developed what has later become known as the "puddle theory," as follows:

> Imagine a puddle waking up one morning and thinking, "This is an interesting world I find myself in, an interesting hole I find myself in, fits me rather neatly, doesn't it? In fact, it fits me staggeringly well, must have been made to have me in it!" This is such a powerful idea that as the sun rises in the sky and the air heats up and as, gradually, the puddle gets smaller and smaller, it's still frantically hanging on to the notion that everything's going to be all right, because this World was meant to have him in it, was built to have him in it; so the moment he disappears catches him rather by surprise. I think this may be something we need to be on the watch out for.[1]

The English poet Rupert Brooke had another and equally sardonic take on species-centrism, in his darkly comic poem "Heaven," as imagined by fish:

> . . . This life cannot be All, they swear,
> For how unpleasant, if it were!
> One may not doubt that, somehow, Good
> Shall come of Water and of Mud;
> And, sure, the reverent eye must see
> A Purpose in Liquidity.
> We darkly know, by Faith we cry,
> The future is not Wholly Dry.
> Mud unto mud!—Death eddies near—
> Not here the appointed End, not here!
> But somewhere, beyond Space and Time.
> Is wetter water, slimier slime!
> And there (they trust) there swim meth One
> Who swam ere rivers were begun,
> Immense, of fishy form and mind,
> Squamous, omnipotent, and kind;
> And under that Almighty Fin,
> The littlest fish may enter in.
> Oh! never fly conceals a hook,
> Fish say, in the Eternal Brook,
> But more than mundane weeds are there,
> And mud, celestially fair;
> Fat caterpillars drift around,
> And Paradisal grubs are found;
> Unfading moths, immortal flies,
> And the worm that never dies.
> And in that Heaven of all their wish,
> There shall be no more land, say fish.

But perhaps we should be somewhat more serious.

The anthropic principle was first introduced, it appears, by the astrophysicist Brandon Carter at a conference in Krakow, Poland, celebrating the five hundredth anniversary of the birth of Copernicus. The venue is, in a sense, ironic, given that Copernicus helped evict the Earth—and thus, humanity—from its prior centrality, while the anthropic principle threatens (or promises) to reestablish this centrality. For Carter, "our location in the universe is *necessarily* privileged to the extent of being compatible with our existence as observers." Here, "location" means not just our physical coordinates in space but also our existence within particular intervals of time.

Before Brandon Carter, Alfred Russell Wallace (codiscoverer with Darwin of the principle of natural selection) seems to have anticipated the anthropic principle in 1904, when he wrote, "Such a vast and complex universe as that which we know exists around us may have been absolutely required . . . in order to produce a world that should be precisely adapted in every detail for the orderly development of life culminating in man."[2]

In *A Brief History of Time*, Stephen Hawking described a number of physical constants and astrophysical phenomena that seem at least consistent with the anthropic principle, including such questions as "Why did the universe start out with so nearly the critical rate of expansion that separates models that recollapse from those that go on expanding forever, such that even now, ten thousand million years later, it is still expanding at nearly the critical rate?" Hawking explains, "if the rate of expansion one second after the Big Bang had been smaller by even one part in a hundred thousand million million, the universe would have recollapsed before it ever reached its present size."[3] In short, we would have been victimized by a kind of Big Crunch.

Time, now, for a brief excursion to the "cosmological constant," introduced by Albert Einstein, and which he considered his "greatest blunder"—but which now seems remarkably prescient. Einstein was troubled by the fact that gravity would cause the universe to collapse onto itself (that Big Crunch), so he introduced a "constant," essentially out of thin air, that pulled in the opposite direction, causing the cosmos to remain stable. Bear in mind that Einstein had been working before Edwin Hubble discovered that the universe was actually expanding. Today, the cosmological constant is widely thought to be intimately connected to so-called dark energy, and physicists such as Steven Weinberg—not a religious believer—point out that if this constant were just a smidgeon larger, then instead of a Big Crunch, the universe would be vaporously insubstantial, expanding at a rate that precludes the formation of galaxies, never mind planets.

Devotees of the anthropic principle have yet more ammunition. Thus, after Wallace but before Carter, the physicist Robert Dicke[4] noted in 1961 that the age of the universe (currently estimated at 14.5 billion years) reflects a kind of Goldilocks Principle, a "golden interval" in which it is neither too young nor too old, but just right. If the universe were younger—that is, if the Big Bang had occurred in the more recent past—it would not have allowed for enough time to accumulate elements heavier than hydrogen and helium via nucleosynthesis. There also wouldn't be any medium-size, rocky planets and thus, no us. By the same token, if the universe were substantially older than it is, nearly all stars would be too elderly to remain part of what astrophysicists call the "main sequence," having matured into white and red dwarfs. As a result, there wouldn't be any stable planetary systems. And once more, no us.

A similar argument can be raised concerning the four fundamental interactions connecting mass and energy: gravitation, electromagnetic attraction and repulsion, and the "strong" and "weak" nuclear forces. These can be seen as balanced in precisely the manner needed to produce matter, and thus, ultimately, the emergence of life. The strong interaction is what holds neutrons and protons together in an atomic nucleus, and that also binds quarks together to form the various subatomic particles. If this strong force were just a tiny bit stronger, then nuclear fusion would have converted the universe's hydrogen into helium, and water—essential for life as we know it—wouldn't exist.

There are other perspectives. For example, the physicist Fred Adams maintains that the necessary conditions for a life-supporting universe aren't so demanding after all. "The parameters of our universe," he writes, "could have varied by large factors and still allowed for working stars and potentially habitable planets."[5] Sure enough, in February 2017, NASA astronomers excitedly announced that they had discovered seven Earth-sized planets orbiting a dwarf star, three of which appear to be in the "habitable zone," including a reasonable likelihood of liquid water. This system, known as Trappist-1, is about forty light years from Earth, and there is every reason to believe that the basic laws of physics obtain there as they do here.

Nonetheless, Adams noted,

> The force of gravity could have been 1,000 times stronger or 1 billion times weaker, and stars would still function as long-lived nuclear burning engines. The electromagnetic force could have been stronger or weaker by factors of 100. Nuclear reaction rates could have varied over many orders of magnitude. Alternative stellar physics could have produced the heavy elements that make up the basic raw material for planets and people. Clearly, the parameters that determine stellar structure and evolution are not overly fine-tuned.[6]

What to believe? The physics involved quickly becomes incomprehensible (at least to me!). For example, Ulf-G Meißner, chair in theoretical nuclear physics at the Helmholtz Institute, University of Bonn, authored a study titled "Anthropic Considerations in Nuclear Physics," published in the journal *Science Bulletin*. Seeking evidence on either side of the anthropic principle, Professor Meißner used supercomputers to simulate worlds in which the fundamental parameters of particle physics and astrophysics were varied, seeking to ascertain

> how sensitive the generation of the light elements in the Big Bang is to changes in the light quark mass m_q and also, how robust the resonance condition in the triple alpha process, i.e. the closeness of the so-called Hoyle state to the

energy of 4He+8Be, is under variations in m_q and the electromagnetic fine structure constant α_{EM}.[7]

Duly noted.

The anthropic principle exists in two primary forms, often designated "strong" and "weak." Oversimplifying, the weak principle is teleological, asserting that whatever conditions are observed in the universe must allow the observer to exist. In short, if these constants weren't as they are, we wouldn't be around to worry about them. The physicist Roger Penrose explained the weak form as follows:

> The argument can be used to explain why the conditions happen to be just right for the existence of (intelligent) life on the Earth at the present time. For if they were not just right, then we should not have found ourselves to be here now, but somewhere else, at some other appropriate time . . . . At any other epoch, so the argument ran, there would be no intelligent life around in order to measure the physical constants in question—so the coincidence had to hold, simply because there would be intelligent life around only at the particular time that the coincidence did hold![8]

To this, Stephen Hawking adds that even slight alterations in the life-enabling constants of fundamental physics in this hypothesized multiverse could "give rise to universes that, although they might be very beautiful, would contain no one able to wonder at that beauty."[9]

Whereas the weak version of the anthropic principle poses a logical conundrum, the strong version is essentially a statement of religious belief, namely that the universe has been tuned as it is because this was required—presumably by some divine creator—in order to establish conditions for human life (and/or perhaps those hairy nosed wombats). An even stronger version has been called the "fixed anthropic principle,"[10] namely that "intelligent information-processing must come into existence in the Universe, and, once it comes into existence, will never die out." Martin Gardner dubbed it the "completely ridiculous anthropic principle" (CRAP).[11]

There are some interesting nonsardonic responses to the anthropic assertion, whether in its weak, strong, or fixed version. Thus, maybe the various physical constants only *appear* to have been organized with carbon-based life in mind, whereas given the inherent nature of mass and energy, they are simply the only values that they could possibly have. In short, are there any "free parameters" in physics? "What really interests me," wrote Einstein, "is whether God had any choice in the creation of the world."[12] Note that Einstein explicitly denied belief in a traditional sky-god; instead, he used "god talk" as a scientific updating of Spinoza, referring to the manifold and complex characteristics of the physical universe. Whether "God had any choice" was his way of asking whether such things as the speed of light, the charge of the electron, proton, and so forth, are fixed or susceptible to alternatives.

Isaac Newton had been troubled that according to his own recently enunciated law of gravity, the solar system should eventually become unstable. Although no slouch when it came to scientific insights, Newton came to believe that God had to intervene periodically to keep things on track, which is to say, He signed on to a kind of eternal service warranty, just to keep our world—and thus, ourselves—viable. (It appears more than peculiar that God couldn't have orchestrated a system that didn't require such house calls. Modern physicists are largely agreed that the problem is solved by counting the complex gravitational influence of the solar system's planets and moons, whose combined effects were too complex for Newton to calculate.)

The question, in any event, is whether reality as we know it is in some sense "preordained"—but by the laws of physics rather than by some omnibenevolent and omnipotent creator. At present, we simply don't know whether it would be equally possible, in some other universes, for entirely different laws of physics to obtain, or whether the current astronomical orientation as well as the different physical and energetic constants are independently derived by the action of a divine dial-twiddler, or whether, if and when physicists eventually come up with a grand unifying Theory of Everything, it may turn out that the various laws and physical constants are somehow bound together, such that the way the world works isn't the result of an external, human-affirming providence but rather, the only way it could, given the nature of matter and energy.

Among the philosophical aspects of the weak anthropic principle (as distinct from its scientific components) is whether the issue is essentially a tautology: if something—anything—is observed about the universe, then if nothing else, that universe must consist of characteristics that permit the observer to exist. This, in turn, is reminiscent of Descartes's *cogito ergo sum*: "I think, therefore I am." But even more so, perhaps, of Ambrose Bierce's *cogito cogito ergo cogito sum* ("I think I think, therefore I think I am")[13] which, Bierce playfully suggested, was about as close to truth as philosophy was likely to get.

Here is another truth: the universe is a big place, and nearly all of it wouldn't permit life—at least, not the kind of carbon-based, water dependent life of which we partake. This, in turn, contributes to the temptation to conclude that our very existence is evidence for a beneficent designer. After all, if we were randomly placed in the universe, we would be dead almost instantly. But of course, we aren't randomly located in the cosmos. Given the abundance of other possible locations, if we existed simply as a result of chance alone, we'd find ourselves (very briefly) somewhere in the very cold empty void of outer space. But we're not the outcome of a purely random process: we happen to find ourselves on the third planet from the sun, a place that has sufficient oxygen, liquid water, moderate temperatures, and so forth . . . just what we need. Looking at ourselves this way, it isn't merely coincidental that we occupy a planet that is suitable for life, because we couldn't possibly be living some place that wasn't!

By the same token, it isn't amazing that the Earth is not a hot gas giant, because if it were, we wouldn't be congratulating ourselves on the fact that we haven't been melted or vaporized. Nor is it remarkable that no matter how tall or short a person may be, her legs are always precisely long enough to reach the ground. One might also turn things around and ask why the universe was "created" to be so *inhospitable* to life. After all, it contains stars, asteroids, comets and galaxies beyond count that are very hot, very cold, highly radioactive, and so forth, and therefore devoid of anything identifiable as "life." So instead of evincing supernatural benevolence, maybe those silent, infinite spaces that unsettled Pascal are evidence of cosmic malevolence or—more likely—indifference.

As for the location of planet Earth within that universe, the British cosmologist and astrophysicist Martin Rees puts it this way:

> We find ourselves on a planet with an atmosphere, orbiting at a particular distance from its parent star, even though this is a very "special" atypical place. A randomly chosen location in space would be far from any star—indeed, it would be in the intergalactic void millions of light-years from the nearest galaxy.[14]

THEN THERE IS another question—not necessarily deeper, but for some thinkers, more perplexing. More than three centuries ago, in *The Principles of Nature and Grace, Based on Reason*, Gottfried Leibniz (1714) noted, "the first question which we have a right to ask will be, 'Why is there something rather than nothing?' "[15] (Here, I'm especially fond of Sidney Morgenbesser's reply to Leibnitz, that "If there were nothing, you'd still be complaining!") But there *is* something, and of course, if there weren't, there wouldn't be any opportunities for complaint. This is not only a genuine, no-nonsense answer to the question of whether the universe has been fine-tuned for us; it also points to a more general merging of statistics, logic, and common sense, namely the difference between probabilities before and after an event.

For example, the philosopher Niall Shanks[16] asks us to imagine shuffling a deck of cards and then dealing them out, face down. What is the likelihood that someone were to predict the entire sequence, in advance, and without any hanky-panky? The chance of getting the first card correct is 1 in 52. The chance of getting the first two cards correct is $1/52 \times 1/51 = 1/2652$, and so on, so that the probability of guessing the entire deck in the proper order is $1/52!$.[a] This is an unimaginably small number, something

---

[a] The notation "!" in mathematics is described as "factorial," with $52! = 52 \times 51 \times 50 \ldots \times 1$.

like one in ten followed by sixty zeros. And yet, it is also true that the chance of the cards having been dealt in the order that actually occurred is 100%. They had to come out *some* way, and among the near infinity of possibilities, one in particular actually emerged. Is that astounding? Yes, if you concern yourself with the chance of that precise outcome, *before* it happened. But no, if you look at the post hoc outcome, knowing that it had to be one way or another.

Alternatively, consider the probabilities before versus after a simple physical event, such as the position of a golf ball before as compared to after a golfer hits it. It would take a near miracle to identify, in advance, the precise eventual location of that ball. But the outcome—wherever the ball ends up—isn't miraculous at all, nor is it evidence of divine intervention or of the golf course having been designed so as to guarantee that particular eventual placement of the ball, since it had to be somewhere. Even though any one specific location is extraordinarily unlikely, it is even less likely that the ball disappeared entirely or that in ended up in the mouth of a hippopotamus. For us to marvel at the fact of our existing (in a universe that permits that existence) is comparable to a golf ball being astounded at the fact that it ended up somewhere.

There are many ways to interpret what might be called the unexpectedness of our existence, none of which necessarily supports the claim that we must attribute that existence to particular preplanning, in our favor, by the cosmos. Every person exists because a particular egg (one out of roughly five hundred ovulated by his or her mother) encountered a particular sperm (one out of roughly 150 million produced by his or her father in a single ejaculation). Every member of the human population— roughly 7.5 billion—can therefore insist that his or her existence was foreordained, evidence of a kind of me-thropic principle.

For a more wide-ranging example, the Chicxulub asteroid crashed into what is today Mexico's Yucatan peninsula 66 million years ago, eventually wiping out the dinosaurs and clearing a path for the rise of mammals and eventually, us. Without that impact and the ecological niches that were opened up by the demise of those previously dominant dinosaurs, it is extremely unlikely that our species would ever have evolved. Had things proceeded differently, the prospect is vanishingly smaller yet that I would be writing this book, or that you would be reading it. Should we therefore see the Chicxulub impact as further evidence that our planet was fine-tuned, with the writing and reading of the present book in mind? And that the dinosaurs' destruction was collateral damage en route to the ultimate goal of creating *Homo sapiens* roughly 65 million years later? If so, then we are responsible for that asteroid and were it not for the universe's goal of producing us, *Tyrannosaurus rex* and company would still be around.

Physics has additional possible explanations for what masquerades as cosmic fine-tuning. Of these, one of the more intriguing (albeit difficult to grasp) is the possibility of "multiverses," which revisits the question of probabilities before versus after an event, albeit in different guise. Here is Martin Rees, once again:

> There may be many "universes" of which ours is just one. In the others, some laws and physical constants would be different. But our universe would not be just a random one. It would belong to an unusual subset that offered a habitat conducive to the emergence of complexity and consciousness. The analogy of the watchmaker would be off the mark. Instead, the cosmos may have something in common with an off-the-rack clothes shop: if the shop has a large stock, we are not surprised to find one suit that fits. Likewise, if our universe is selected from a multiverse, its seemingly designed or fine-tuned features would not be surprising.[17]

Under the multiverse hypothesis, not only is there a potentially infinite number of universes but also the basic physical laws and constants might well vary across them. It is a radically difficult concept, but perhaps no weirder than basic quantum mechanics, which we now know to be valid. It was recently reported that there are something like two trillion galaxies in the currently known universe, which is about twenty times more than had previously been thought.[18] Each galaxy consists of millions—in some cases, billions—of stars, many of which have their own planets. And although it appears that the fundamental physical constants hold across the known galaxies, the mere fact that there are so many (the overwhelming majority of which are not in any meaningful sense "known") opens the possibility that our Earthly experience may be a small subset of the possible—even without introducing the prospect of multiple universes.

Niall Shanks suggests that the multiverse hypothesis "does to the anthropic universe what Copernicus's heliocentric hypothesis did to the cosmological vision of the Earth as a fixed center of the universe."[19] Now, post-Copernicus (and Kepler, Galileo, and others), the Earth is known to be just one planet among many, in one galaxy among many. Perhaps we're just the occupants of one universe among many. Interestingly, even as he demoted the Earth, Copernicus himself placed the Sun in the center of the universe, just as he assumed that planetary orbits were precise circles, an assumption that was widespread in early astronomy, based on the notion that the "heavenly bodies" are necessarily perfect, just as, in their geometry, circles are perfect. Galileo, too, assumed that planetary orbits were circular; it wasn't until Kepler—using data from the aforementioned Tycho Brahe—that astronomers recognized that they are elliptical. The cosmos, like the human body, is far from perfect. But like the human body and the bodies of all other organisms, it is good enough to have permitted us and them to exist.

For extraterrestrial life to exist, it seems likely (although by no means certain) that it would have to reside on one or more exoplanets, asteroids, or perhaps a comet, rather than within a star or free-floating in open space. Moreover, such exoplanets would have to be associated with stars that, for example, don't emit massive amounts of X-rays or other forms of radiation. But of course, this presumes that "life" would be consistent with "life as we know it." Maybe there are critters out there who cheerfully bathe in hefty levels of what to terrestrial biologists are lethal amounts of energy, or get by and even thrive on not enough energy to sustain a perseverating entity that would qualify—to us—as "alive."

Just a bit closer to established reality, quantum mechanics offers another potential solution to the anthropic conundrum, one that seems if anything weirder than the multiverse hypothesis. According to theory—the same theory that gives rise to, among other things, the very real computer on which this book has been written—matter at its most fundamental level is made up of probabilistic wave functions, which only transition to "reality" when a conscious observer intervenes to measure or perceive them. In the famous "double slit experiment," light is revealed to be either a particle or a wave only *after* it is measured as one or the other. Prior to this, photons do not, in a sense, exist as clear-cut entities; afterward, they do.

Based on these and other findings, the physicist John Wheeler, one of the towering pioneers of quantum mechanics, who coined the term "black hole" (and who numbered the Nobel laureate Richard Feynman among his students) suggested a "participatory anthropic principle," whereby—believe it or not—the universe had to include conscious beings in order for it to exist. Personally, I don't believe it.

I do, however, believe in evolution, which leads—by a stretch no more bizarre than Wheeler's—to the suggestion that maybe it shouldn't be surprising that we live in a universe suitable for life, something that has happened not because the universe has been fine-tuned for us or has somehow been "made real" by us, but because *we* are fine-attuned to *it* because of natural selection. Just as the physical qualities of air have selected for the structure of bird wings, and the anatomy of fish speaks eloquently about the nature of water, maybe the nature of the physical universe has in the most general sense, selected for life, and thus, for us.

There is also another, more peculiar way of incorporating natural selection into the anthropic quest. What if natural selection occurs at the level of galaxies, or even universes, such that those offering the potential for life are more likely to replicate themselves? If so, then compared to life-denying galaxies, life-friendly ones might conceivably have produced more copies of themselves, providing greater opportunities for life forms such as ourselves. Aside from the rampant unlikelihood of this "explanation," it remains unclear how or why such pro-life galaxies would be favored over their more barren alternatives. This would be necessary for natural selection to take place at the level of galaxies, or universes.

Nonetheless, the physicist Lee Smolin has pursued the notion of "cosmological natural selection," whereby perhaps not just galaxies but entire universes replicate themselves, courtesy of black holes.[20] If so, then what sort of universes would be favored—"selected for," as biologists would put it? Easy: those that employ physical laws and constants that are "more fit," that is, that lend themselves to being reproduced. This conveniently explains (if explanation is the correct word) why our universe contains black holes (it's how the ones that replicated managed to do so). It also leads to the supposition that perhaps intelligent beings can contribute to the selective advantage of their particular universe, via the production of black holes, and who-knows-what-else.

When it comes to the physicists' concept of dark and mysterious things, we'd be remiss to ignore dark matter and its potential, inscrutable bestiary. Thus, astrophysicists are pretty much agreed that normal matter contains only about one-fifth the energy contained in its obscure cosmic doppelganger. What, then, is our basis for assuming that normal stuff is the only stuff that counts? And the only stuff that harbors life? The MIT astrophysicist Lisa Randall proposed that "dark life could in principle be present—even right under our noses. But without stronger interactions with the matter of our world, it can be partying or fighting or active or inert and we would never know."[21]

Perhaps here we should take a cue from the philosopher Ludwig Wittgenstein, in his prodigiously titled *Tractatus Logico-Philosophicus*: "What we cannot speak about, we must pass over in silence."[22] At least for now.

Yet another possibility, however, no less weird that living dark matter, was broached by Carl Sagan in his 1985 novel *Contact*. In it, the heroine is advised by an extraterrestrial intelligence to study transcendental numbers—numbers that are not algebraic—of which the best-known example is pi. She computes one such number out to 10^{20} places, at which point she detects a message embedded in this number. Since such numerology is fundamental to mathematics itself and is thus, in a sense, a property of the basic fabric of the universe, the implication is that the cosmos itself is somehow a product of intelligence, since the message is clearly an artificial one and not the result of random noise. Or maybe the universe itself is "alive," and the various physical and mathematical constants are part of its metabolism. Such speculation is great fun, but please bear in mind that it is science *fiction*, not science!

It should be clear at this point that when it isn't verging on unintelligibly abstruse or downright goofy, the anthropic argument readily devolves into speculative philosophy and even theology. Indeed, it is reminiscent of the "god of the gaps" perspective, in which god is posited whenever science hasn't (yet) provided an answer. Calling on god specifically when there is a gap in our scientific understanding may be tempting, but it is not even popular among theologians, because as science grows, the gaps—and thus, god—shrinks. It remains to be seen whether the anthropic principle, in whatever form, succeeds in expanding our sense of ourselves beyond that illuminated by science. I wouldn't bet on it.

As usual, Richard Dawkins said it well: "The universe we observe has precisely the properties we should expect if there is, at bottom, no design, no purpose, no evil and no good, nothing but blind pitiless indifference."[23] And yet, despite what has been called Copernican Mediocrity—to which I would add Darwinian Mediocrity—just because the universe is unlikely to be what it is simply for our benefit, this need not, and should not, give rise to an alternative, "misanthropic principle." Regardless of how special we are, or aren't, wouldn't we be well advised to treat everyone (including the other life forms with which we share this planet), as the precious beings we like to imagine us all to be?

Old paradigm: The universe has been "fine-tuned" for life, especially human life.

New paradigm: There are many alternative explanations for this apparent fact, all of them based on a mechanistic rather than theistic conception of reality.

NOTES

1. Douglas Adams, *The Salmon of Doubt* (New York: Ballantine, 2003).
2. Alfred Russell Wallace, *Man's Place in the Universe: A Study of the Results of Scientific Research in Relation to the Unity or Plurality of Worlds*, 4th ed. (London: George Bell & Sons, 1904).
3. Stephen Hawking, *A Brief History of Time* (New York: Bantam, 1990).
4. Robert H. Dicke, "Dirac's Cosmology and Mach's Principle," *Nature* 192, no. 4801 (1961): 440–441.
5. Fred Adams, 2017, http://cosmos.nautil.us/feature/113/the-not-so-fine-tuning-of-the-universe.
6. Ibid.
7. Ulf-G Meißner, "Anthropic Considerations in Nuclear Physics," *Science Bulletin* 60, no. 1 (2015): 43–54. link.springer.com/article/10.1007%2Fs11434-014-0670-2.
8. Roger Penrose, *The Emperor's New Mind* (London: Oxford University Press, 1989).
9. Stephen Hawking, *A Brief History of Time* (New York: Bantam, 1989).
10. See, for example, John D. Barrow and Frank J. Tipler, *The Anthropic Cosmological Principle* (New York: Oxford University Press, 1988).
11. Martin Gardner, "WAP, SAP, PAP, and FAP," *New York Review of Books* 23, no. 8 (May 8, 1986): 22–25.
12. Quoted in A. Calaprice, *The Expanded Quotable Einstein* (Princeton, NJ: Princeton University Press, 2000).
13. Ambrose Bierce, *The Devil's Dictionary* (Athens: University of Georgia Press, 2000).
14. Martin Rees, *Just Six Numbers: The Deep Forces That Shape the Universe* (New York: Basic Books, 1999).
15. Gottfried Leibniz, *The Principles of Nature and Grace, Based on Reason* (1714).
16. Niall Shanks, *God, the Devil, and Darwin* (New York: Oxford University Press, 2004).
17. Martin Rees, *Our Cosmic Habitat* (Princeton, NJ: Princeton University Press, 2001).

18. See http://www.nature.com/news/universe-has-ten-times-more-galaxies-than-researchers-thought-1.20809.

19. Niall Shanks, *God, the Devil, and Darwin* (New York: Oxford University Press, 2004).

20. Lee Smolin, *The Life of the Cosmos* (New York: Oxford University Press, 1999).

21. Lisa Randall, *Dark Matter and the Dinosaurs* (New York: Ecco, 2015).

22. L. Wittgenstein, *Tractatus Logico-Philosophicus* (New York: Routledge & Kegan Paul, 1961).

23. Richard Dawkins, *The Blind Watchmaker* (New York: Norton, 1986).

6

Tardigrades, Trisolarans, and the Toughness of Life

ALBERT SCHWEITZER WAS, by all accounts, one of the giants of the twentieth century. A talented musician and theologian, Schweitzer attended medical school in his thirties, giving up a much-admired professorship in the humanities, after which he opened a hospital in a remote part of today's Gabon, where he labored selflessly to provide medical treatment for some of the planet's most depauperate and underserved people.[1]

Schweitzer recounted an important moment, floating on a river in Gabon, when his personal philosophy became crystallized:

> Lost in thought, I sat on deck of the barge, struggling to find the elementary and universal concept of the ethical that I had not discovered in any philosophy. I covered sheet after sheet with disconnected sentences merely to concentrate on the problem. Two days passed. Late on the third day, at the very moment when, at sunset, we were making our way through a herd of hippopotamuses, there flashed upon my mind, unforeseen and unsought, the phrase: "Reverence for Life." The iron door had yielded. The path in the thicket had become visible. Now I had found my way to the principle in which affirmation of the world and ethics are joined together.[2]

"Reverence for life" is indeed both an affirmation of the world and an admirable keystone for ethics, directed toward all life, and not merely other human beings. "Just as our own existence is significant to each of us," wrote Schweitzer, "a creature's existence is significant to it."[3] Reminiscing on his early childhood, Schweitzer recalls,

> As far back as I can remember I was saddened by the amount of misery I saw in the world around me . . . . One thing that especially saddened me was that the unfortunate animals had to suffer so much pain and misery . . . . It was quite incomprehensible to me—this was before I began going to school—why in my evening prayers I should pray for human beings only. So when my mother had

prayed with me and had kissed me good-night, I used to add silently a prayer that I composed myself for all living creatures. It ran thus: "O heavenly Father, protect and bless all things that have breath guard them from all evil, and let them sleep in in peace."[4]

One time, as a young boy, Schweitzer went fishing with some friends, but it

was soon made impossible for me by the treatment of the worms that were put on the hook . . . and the wrenching of the mouths of the fishes that were caught. I gave it up . . . .. From experiences like these, which moved my heart . . . there slowly grew up in me an unshakeable conviction that we have no right to inflict suffering and death on another living creature, and that we ought all of us to feel what a horrible thing it is to cause suffering and death.[5]

Reverence for life more than justifies restraint when it comes to inflicting pain and death on other living things. It is also consistent with a profound valuation of life (albeit often restricted to human life only), as explicitly developed in several of humanity's wisdom traditions. Consider this, from Judaism: "Whoever destroys a soul, it is considered as if he destroyed an entire world. Whoever saves a life, it is considered as if he saved an entire world" (Mishnah Sanhedrin 4:5). And from the *Qur'an* Sura 5032 we read, "If any one slew a person . . . it would be as if he slew the whole people; and if any one saved a life, it would be as if he saved the life of the whole people."

Few would argue with the proposition that life—each life—is precious, although perhaps not infinitely so. Life is definitionally distinct from nonlife in many ways, such as responsiveness to stimuli and the capacity to reproduce—with a bottom line being the maintenance of internal conditions that are highly nonrandom and low in entropy. Unlike, say, crystals or salt solutions, life as we know it can only exist within generally narrow limits, with specified concentrations of nutrient molecules, oxygen, and carbon dioxide as well as waste products. Living things, moreover, can only tolerate a narrow range of acid-base balance (pH), of ambient pressure and temperature, of the osmotic concentration of various electrolytes, and so forth. Individual lives aren't only precious but also delicate, often painfully so.

The demanding balance required by a living organism is typically achieved by homeostasis, an array of thermostatic control mechanisms that—like a temperature thermostat in a house—increases something if it gets too low, and decreases it if it gets too high. In his now-classic text, *The Wisdom of the Body*, the physiologist William Cannon detailed the many ways that life maintains itself within a narrow range of parameters, testimony to a pair of conflicting realities, one being that individual lives are delicate in that even small deviations in conditions—especially when it comes to an organism's internal environment—can be lethal. The other is the contradictory fact that by virtue of having the capacity to maintain such narrow limits despite variations in the external environment as well as the dynamic nature of the internal (e.g., the unavoidable accumulation of waste products), life is perseverative. Homeostasis makes organisms capable of colonizing a wide range of environments—which wouldn't be possible if their innards were limited to reflecting their immediate surroundings. Turtles and snails carry their protective houses on their backs; living things are obliged to maintain their internal houses within narrow limits and, accordingly, are able to do so.

In his book *Catastrophe Theory*, the Russian mathematician Vladimir Arnol'd described the "fragility of good things," based on the fact that to be good, a thing will likely require success in a number of different domains, while incapacity in any one leads to failure of the whole. According to Arnol'd, when it comes to stable systems,

> a small change of the parameters is more likely to send the system into an unstable region than into a stable region. This is a manifestation of a general principle stating that all good things (e.g. stability) are more fragile than bad things. It seems that in good situations a number of requirements must hold simultaneously, while to call a situation bad even one failure suffices.[6]

At first blush, Arnol'd's principle appears not only theoretically well grounded (consistent as it is with the Second Law of Thermodynamics), but also productively insightful. But if Arnold's theory itself is a "good thing," is it fragile too?

When it comes to ecological systems, the theory is mostly disconfirmed. Well-functioning ecosystems are generally admired as good, with ecologists agreeing that complex ecosystems are preferable to monocultures. Compare, for example, a magnificently diverse tropical rainforest with a biologically barren and uniformly boring palm oil plantation. The former teems with a fabulous array of interdigitating life forms and is also capable of resisting most external shocks short of outright human-imposed devastation, whereas the latter resembles an investor's portfolio that is undiversified and is thus vulnerable to imported diseases, climate change, and so forth. The point is that when it comes to ecosystems, good things—that is, complex and biodiverse ones—are *more* stable, and thus *less* fragile, than their simpler and biologically worse counterparts.

But such cases may well be exceptional and one needn't be a modern biologist to recognize that good things really are fragile, most of the time. Seneca put it thus: "It would be some consolation for the feebleness of ourselves and our works if all things should perish as slowly as they come into being; but as it is, increases are of sluggish growth, while the way to ruin is rapid." Aristotle, in his *Nicomachean Ethics*, made a similar point: it is possible to fail in many different domains, whereas the path to success is narrow. There are numerous ways, for example, for an archer to miss her mark, but only one way to hit it. A last-minute tremor, gust of wind, broken string, slightly twisted bow, inadequately feathered arrow, a sudden cough: any of these can make the whole process go awry. By the same token, there are many ways for a biological system to fail—and thus, to sicken and die—but less tolerance when it comes to maintaining the demanding conditions necessary for life. Indeed, life can be defined as a concatenation of highly nonrandom, carefully circumscribed events that must all come together to succeed in keeping random, bad things (like death) at bay. As someone first noted somewhere—I believe it was Richard Dawkins, but maybe it was me—there are many more ways to be dead than to be alive.

This would suggest that living organisms are the poster children for fragility. After all, life is almost by definition a circumstance in which entropy is (temporarily) held at bay, through the expenditure of substantial amounts of energy, with everything maintained within narrowly acceptable physiological and anatomic windows. Not enough calories (or too much, or of the wrong sort), inopportune pH or osmotic balance, inadequate titration between micronutrients, and the whole structure falls apart.

Which brings us (finally!) to the subject of the present chapter: recent findings that unlike individual lives, life itself is remarkably robust. It is an important insight, consistent with the underlying message of this book, namely that although life is indeed special, it is not *that* special, just as even though human beings are also a particular case (species *Homo sapiens*) of a more general phenomenon (life itself), we are not *that* special. The fact that we exist, therefore, isn't a "miracle" in itself, because although life is wonderful and extraordinary, not to mention precious and deserving of reverence, it isn't in any sense miraculous.

ENTER: EXTREMOPHILES. THIS word didn't exist a few decades ago, and has only found widespread use in the twenty-first century. It refers to organisms that survive—even, thrive—in environments that are extremely hot, cold, highly acidic or alkaline, and so forth, circumstances that would be lethal for most living things. Not surprisingly, extremophiles tend to be relatively simple creatures, notably invertebrates and especially bacteria, although there is no bright line distinguishing, say, mammals such as arctic hares—which thrive in very cold habitats—from their close relatives such as more "traditional" rabbits, which inhabit temperate environments. In a sense, the former are extremophiles relative to the latter, but neither compares with those life forms whose mere existence has excited the admiration and wonder of biologists.

The first asphalt roads in such eastern US cities as New York, Baltimore, and Washington, DC, were paved with natural asphalt taken directly from Pitch Lake, in Trinidad. The first Europeans to discover this natural curiosity, in 1595, were under the command of Sir Walter Raleigh, who used pitch from this lake to caulk his ship, noting in his diary that the material was "most excellent . . .. It melteth not with the sun as does the pitch of Norway."[7] More than four centuries later, scientists discovered abundant microbial life luxuriating in this same lake of liquid asphalt,[8] an environment that would seem more appropriate to one of Dante's infernal regions than a life-affirming petri dish. Nor is this finding unique.

In 2013 bacteria were reported to be abundant in a cold, dark lake (of more traditional water), one-half mile under the Antarctic ice.[9] A month later, microbes were found occupying the deepest place on Earth, the Marianas Trench.[10] There are also "infra-terrestrials" which live, incredibly, inside rocks in the deep ocean, nearly 2,000 feet below the sea floor, which itself is under 8,500 feet of ocean, and thus not just utterly dark and devastatingly cold, but subject to immense pressure.[11] (This, too, was reported early in 2013, which qualifies as an "annus mirabilis" for the discovery of extremophiles.) Living things—thermophiles—have also been found in superheated oceanic vents, at temperatures up to 122 degrees Celsius, which is considerably hotter than boiling water.

It remains unknown exactly how life first appeared on Earth, not least because it occurred more than 4 billion years ago and left no currently discernible fossil evidence. There have, however, been numerous hypotheses and no shortage of feasible speculation, including famous research by Stanley Miller in the 1950s, which showed that complex organic molecules can arise when simpler chemicals are recirculated in a laboratory and provided with electrical energy (simulating lightning storms). It had long been assumed, especially by nonscientists, that the details of how life first evolved constitute one of the greatest evolutionary mysteries, perhaps *the* greatest. In fact, that's not the case. Although it would certainly be interesting to unravel precisely how nucleic acids (whether DNA, RNA, or some precursor) first achieved the ability to self-replicate, and also how cells first appeared in the pre-Cambrian, the truth is that there are lots of feasible routes for both, and there are many other, far more challenging evolutionary mysteries.[12]

Moreover, the abundance of extremophile life forms and the subsequent recognition that life is resilient and widespread has if anything helped undercut the myth that aliveness is so special that its appearance is not only a huge missing piece of the evolutionary puzzle, but also prima facie evidence for divine intervention, a belief that was widespread—even among many scientists—as recently as the early nineteenth century. The supposed supernatural specialness of life was encapsulated in the doctrine of vitalism, which proclaimed that living things contained some sort of metaphysical "life spark," which was not subject to the basic laws of physics and chemistry. In addition to its metaphysical and theological appeal, evidence for vitalism came from the assumption that unlike inorganic compounds such as salts, oxides, and simple acids and bases, organic compounds (including proteins and carbohydrates) were imbued with a unique vitality such that they not only characterized living organisms but also could only be produced in plant or animal bodies and could not be synthesized in the barren confines of a merely inorganic laboratory.

In 1773, a French chemist isolated crystals of urea from the urine of different animals, as well as from human beings. Even when it turned out that this substance was a comparatively simple compound, vitalism was not undermined, since the greatest chemists maintained that urea could only be made by animals with kidneys. ("Organic" compounds were well known at the time, their name deriving from the dogma that they existed only by virtue of the action of appropriate living organs.) But since previously identified organic compounds were complex and urea was relatively simple, the prospect arose that perhaps it could be artificially synthesized after all.

Then, in 1828, Friedrich Wöhler—who was already renowned as the discoverer of aluminum—undertook the synthesis of the inorganic compound, ammonium cyanate. What he got, to his surprise and consternation, was urea: surprise because he wasn't expecting this result, and consternation because, at the time, Wöhler himself was a firm believer in vitalism. He wrote to his mentor, the famous chemist Jacob Berzelius, "I cannot, so to say, hold my chemical water and must tell you that I can make urea without thereby needing to have kidneys, or anyhow, an animal, be it human or dog." It was no longer true that bona fide organic compounds could only be made by living organisms and then extracted from biological sources such as blood, urine, and so forth. This led Wöhler to write additionally to Berzelius, lamenting "the great tragedy of science, the slaying of a beautiful hypothesis by an ugly fact."[13]

From today's perspective, it isn't at all clear that the chemical synthesis of organic compounds is ugly—although it is certainly a fact, one that is replicated innumerable times throughout the world and has given birth to the field of organic chemistry, which currently synthesizes about a million different compounds each year, including many hormones, vitamins, and life-saving antibiotics. But the hypothesis of vitalism—whether or not beautiful—is dead as can be. The success of organic chemistry thus contributed to the demise of yet another paradigm: that life itself, and not merely its human component, was metaphysically unique. (The notion nonetheless persisted in some circles of thought—humanist rather than biological. Thus, the philosopher Henri Bergson argued as late as 1907 that living things were alive by virtue of their "élan vital," which led the biologist Julian Huxley to note that this was as useful as attributing the movement of a railroad train to its "élan locomotif.")

This brings us back to those extremophiles, the discovery of which hasn't been quite as earthshaking and paradigm-breaking as the demolition of vitalism, but which nonetheless added another nail in the coffin that proclaims life to be a creation of God because it couldn't possibly derive from natural processes. Given that organisms can succeed in extreme environments, maybe they first evolved in them as well. It was long assumed, for example, that life must have originated in some sort of warm, shallow, benevolent puddle that offered the kind of comfortable incubator that such a delicate flower would require. This may have been the case. However, the existence of thermophiles thriving in superheated, hydrothermal deep ocean vents, along with the discovery of numerous other extremophiles, has raised the prospect that perhaps life first emerged in what we—sunny children of a relatively easy, superficially life-friendly environment—have until recently considered impossible conditions.

It has also evoked considerable interest from the community of astrobiologists, those scientists interested in ascertaining whether life exists elsewhere in the solar system and/or beyond. Hence, the special significance of not only life abounding in extreme conditions—superheated (thermophiles), supercooled (cryophiles), without oxygen (anaerobes), and intensely salty (halophiles)—but also getting nutrition from methane (methanotrophs) and surviving, even thriving, among heavy concentrations of such traditionally toxic heavy metals as arsenic, cadmium, copper, lead, and zinc. There are even some radiation-resistant organisms that can cheerfully gargle with the effluent from nuclear reactors.

Most extremophiles are microbes, but not all. There are, for example, a group of wingless, mostly eyeless insects known as grylloblattids, more commonly ice bugs or ice crawlers. They live, as you might expect, in very cold environments, typically under frozen rocks. My personal favorites, however, are tardigrades. These multicellular creatures are rarely more than one millimeter in length and often invisible to the unaided eye. They have four legs along each side, each outfitted with tiny claws. They also have a clearly discernible mouth and are simply adorable. Purists don't include tardigrades among extremophiles, since they do not appear adapted to extreme environments per se—that is, like us, they do best in comparatively benign conditions, which, in the case of tardigrades includes the moist, temperate miniworld of forest moss and lichens. Their probability of dying increases in proportion as they are exposed to highly challenging circumstances, so, unlike classic extremophiles, they are evidently adapted to what human beings, at least, consider moderate circumstances.

However, tardigrades are extraordinary in their ability to survive when their environments become extreme. Not only that, but whereas typical extremophiles specialize in going about their lives along one axis of environmental extremity—extreme heat or cold, one or another heavy metal, and so forth—tardigrades can survive when things get dicey along many different and seemingly independent dimensions, simultaneously and come what may. You can boil them, freeze them, dry them, drown them, float them unprotected in space, expose them to radiation, and even deprive them of nourishment—to which they respond by shrinking in size. These creatures, also known as water bears, are featured on appealing T-shirts with the slogan "Live Tiny, Die Never," and a delightful rap song describing their indifference to extreme situations is titled "Water Bear Don't Care." These little creatures might be the toughest on Earth.

You can put them in a laboratory freezer at -80 degrees Celsius, leave them for several years, then thaw them out and just twenty minutes later they'll be dancing about as though nothing had happened. They can even be cooled to just a few degrees above absolute zero, at which atoms virtually stop moving; once thawed out, however, they move around just fine. (Admittedly, they aren't speed demons; the word "tardigrade" means "slow walker.") Exposed to superheated steam (140 degrees centigrade), they shrug it off and keep on living. Not only are tardigrades remarkably resistant to a wide range of what ecologists term environmental "insults" (heat, cold, pressure, radiation, etc.) but also they have a special trick up their sleeves: when things get really challenging—especially if dry or cold—they convert into a spore-like form known as a "tun," which can live (if you call their unique form of suspended animation "living") for decades, possibly even centuries, and thereby survive pretty much anything that nature might throw at them. In this state, their metabolism slows to less than 0.01% of normal.

Given that they possess the kind of powers we otherwise associate with comic book superheroes, it might seem that tardigrades are creatures out of science fiction, but the connection could well be the other way around. In 2015, *The Three Body Problem*, a blockbuster that has broken all records for sci-fi literature in China, became the first book not originally published in English to win the coveted Hugo Award for best sci-fi novel. It describes extraterrestrials known as Trisolarans, whose planet is associated with three suns, the real-life interactions of which—as physicists and mathematicians understand—would generate chaotically unstable conditions. Trisolarans, therefore, are unpredictably subjected to extreme environments depending on the temporary orientation of their planet relative to its chaotically interacting stars: sometimes lethally hot, other times cold, sometimes unbearably dry and bright, other times dark, and so forth. As a result, these imagined extremophiles have evolved the ability to desiccate themselves, rolling up like dried parchment, only to be reconstituted when conditions become more favorable.

I have not been able to determine whether the author Liu Cixin was aware of real-life, Earth-inhabiting tardigrades when he invented his fictional Trisolarans, but the convergence is striking.[a] (In the interest of scientific open-mindedness it should also be considered that perhaps tardigrades are real Trisolarans, refugees from a planet that was chronically exposed to intense environmental perturbations. This would explain the puzzling fact that tardigrades appear hyperadapted, able to survive to extremes that greatly exceed what they experience here on Earth.)

---

[a] My guess is that he wasn't; Mr. Liu is masterful when it comes to physics, but with biology (or, for that matter, psychology) . . . not so much.

In any event, tardigrades have two more arrows in their extremophile quiver, neither of them shared with Mr. Liu's Trisolarans. For one, it has recently been discovered that they possess genes that produce a peculiar array of constituent chemicals, known as "intrinsically disordered proteins," which help induce a solid internal state in these animals when they experience desiccation.[14]

And for another, fully one-sixth of their DNA appears to consist of genetic material from other species,[15] although this finding has been disputed.[16] It has long been known that some cross-species—horizontal—transfer of DNA takes place, but not on the scale exhibited by water bears. Most animals sport less than 1 percent foreign DNA. A possible explanation for how tardigrades developed this trick is that when these animals dehydrate in response to environmental exigency, their DNA breaks into fragments, after which it is reconstituted upon rehydration. (It isn't known whether upon returning to normal, individual tardigrades retain their prior memories and personalities, especially if they have recently absorbed new dollops of DNA—and from different species, yet.)

Seriously, all species possess some capacity to repair errors in their genome; those individuals and hence species lacking this ability will have been selected against, simply given the unavoidable tendency of DNA to mutate, even without the extra challenge of adjusting to extreme conditions. As tardigrade genes reconstitute themselves from the tun state, their cell and nuclear membranes become leaky, whereupon they might have acquired the habit of incorporating various pieces of nontardigrade DNA that are inevitably floating around.

Although presumably the stitching in of extra-specific DNA would be initially random, it is entirely possible that some of these genomic additives were especially likely to contribute to their possessor's fitness, whereupon they would have been projected into future generations, a process made all the more likely under conditions of fluctuating environmental harshness. If so, then if the initial report of extra-specific DNA in these creatures turns out to be accurate, one consequence of tardigrade adaptation to extreme conditions—for example, the ability to close down their metabolism and suspend normal DNA structure—will itself have contributed to their ability to adapt to those conditions, by facilitating the incorporation of those occasional DNA snippets that enhance survival and reproduction.

However this process began and if for whatever reason it persists, a high level of cross-species DNA transfer among tardigrades would suggest a rethinking of the traditional understanding whereby genes are only transmitted vertically, from parents to offspring. So here is yet another paradigm about to be lost, thanks to these unusual extremophiles, whereby a linear tree of life must make room for a complex, horizontally branching web. In his book *What Is Life?*, the renowned mathematical biologist J. B. S. Haldane wrote, "the universe is not only queerer than we imagine but queerer than we *can* imagine."[17] Extremophiles such as tardigrades demonstrate that life is not only "queerer" than most of us have imagined, but also more resilient.

In an increasingly bio-unfriendly world, made all the more so when the United States elected a rabidly antienvironmental president in 2016, extremophiles in general and tardigrades in particular provide a tiny ray of hope. There also seems little doubt that they deserve all the reverence that anyone—and not just Albert Schweitzer—can muster.

At the same time, those extremophiles, with their stubborn life-affirming resilience, are a reminder to bestow on ourselves a hefty dose of humility, along the lines of Ralph Waldo Emerson's poem "Hamatreya," which comments on Earth's "boastful boys . . . Who steer the plough, but cannot steer their feet / Clear of the grave."

Old paradigm: Life is delicate; hence the fact that we are alive is testimony to our profound specialness.

New paradigm: Although individual lives are delicate, life in one form or another is remarkably robust; hence, aliveness isn't in itself a statement of any living thing's extraordinary importance.

NOTES

1. Some material in this chapter has been modified from D. Barash, "Live Tiny, Die Never," *Nautilus* (2017), https://www.scribd.com/article/342232859/Live-Tiny-Die-Never-Behold-The-Toughest-Animal-On-Earth.
2. Albert Schweitzer, *Out of My Life and Thought: An Autobiography*, 60th anniversary ed. (Baltimore, MD: Johns Hopkins University Press, 2009).
3. Albert Schweitzer, *Reverence for Life: The Words of Albert Schweitzer*, comp. Harold E. Robles. (New York: HarperCollins, 1993).
4. Ibid.
5. Ibid.
6. Vladimir Arnol'd, *Catastrophe Theory* (New York: Springer, 2003).
7. Seneca, Lucius Annaeus. *Letters From a Stoic* (Mineola, NY: Dover Publications, 2013).
8. Dirk Schulze-Makuch, Shirin Haque, Marina Resendes de Sousa Antonio, et al., "Microbial Life in a Liquid Asphalt Desert," *Astrobiology* 11, no. 3 (2011): 241–258.
9. James Gorman, "Bacteria Found Deep under Antarctic Ice, Scientists Say," *New York Times*, February 6, 2013.
10. Ronnie Glud et al. "High Rates of Microbial Carbon Turnover in Sediments in the Deepest Oceanic Trench on Earth," *Nature Geoscience* 6, no. 4 (2013): 284–288.
11. Ibid.
12. D. P. Barash, *Homo Mysterious: Evolutionary Puzzles of Human Nature* (New York: Oxford University Press, 2012).
13. Quoted in J. Erik Jorpes, *Jac. Berzelius—his life and work* (Berkeley: University of California Press, 1970).

14. T. C. Boothby et al., "Tardigrades Use Intrinsically Disordered Proteins to Survive Desiccation," *Molecular Cell* 65, no. 6 (2017): 975–984.

15. T. C. Boothby et al. "Evidence for Extensive Horizontal Gene Transfer from the Draft Genome of a Tardigrade," *Proceedings of the National Academy of Sciences,* 112, no. 52 (2015): 15976–15981.

16. G. Koutsovoulos et al., "No Evidence for Extensive Horizontal Gene Transfer in the Genome of the Tardigrade *Hypsibius dujardini,*" *Proceedings of the National Academy of Sciences* 113, no. 18 (2016): 5053–5058.

17. J. B. S. Haldane, *What Is Life?* (London: Boni and Gaer, 1947).

7

Of Humanzees and Chimphumans

NO ONE HAS yet cloned a human being, although the barriers to doing so are not so much scientific or biological as primarily ethical and legal. There is every reason to think that given a serious effort, *Homo sapiens* could be cloned, as has already been done for dogs, cats, sheep, goats, cattle, horses, and so forth. It is a bit more of a stretch—but by no means impossible or even unlikely—that a hybrid or "chimera" (composed of parts derived from two closely related species) combining the genotypes of a human being and a chimpanzee could be produced in a laboratory. After all, human and chimp (or bonobo) share, by some estimates, roughly 99% of their DNA, with the human-gorilla genetic overlap at approximately 98%. Granted that the 1 percent difference in the former case presumably involves some key alleles, the new gene-editing tool CRISPR offers the prospect (for some, the nightmare) of adding and deleting targeted genes as desired. As a result, it is not unreasonable to foresee the possibility—eventually, perhaps, the likelihood—of producing "humanzees" or "chimphumans."

During the 1920s, a Russian biologist with the marvelously Slavic name Ilya Ivanovich Ivanov appears to have made the first serious, scientifically informed efforts to create a genetic hybrid between chimpanzees and human beings. Ivanov had the perfect qualifications: not only did he possess a special interest in creating interspecific hybrids, but he was also an early specialist in artificial insemination and had achieved international renown as a successful pioneer when it came to horse breeding. Prior to his work, even the most prized stallions and mares were limited to reproducing by "natural cover"—that is, the old-fashioned way, one mounting at a time. But Ivanov found that by appropriate and careful dilution of stallion semen, combined with adroit use of the equine equivalent of a turkey baster, he could generate as many as five hundred foals from a single genetically well endowed stallion. His achievement caused a worldwide sensation, but nothing compared to what he next attempted.

And failed.

It happened initially at the Research Institute of Medical Primatology, the oldest primate research center in the world, located at Sukhumi, the capital of Abkhasia, currently a disputed region in the state of Georgia, along the Black Sea. At one time, the Sukhumi Institute was the largest facility conducting research on primates. Not coincidentally, Stalin is believed to have been interested in such efforts, with an eye toward developing the "new Soviet man" (or half-man, or half-woman).

Nor was Soviet interest in combining human and nonhuman genetic material limited to Russian biologists. The novelist M. Bulgakov, best known—at least in the West—for his fantasy *The Master and Margarita*, also wrote *Heart of a Dog*, a biting satire on early Soviet-era social climbers, in which a pituitary gland from a drunken person is implanted into a stray dog, who subsequently becomes more and more human—although not noticeably more humane as he proceeds to eliminate all "vagrant quadrupeds" (cats) from the city. Maxim Gorky was on board, writing approvingly that Lenin and his Bolshevik allies were "producing a most severe scientific experiment on the body of Russia," which would eventually achieve "the modification of human material."[1]

Efforts at cross-species manipulation became a staple of Soviet biology, as when S. A. Voronov attempted "rejuvenation therapy," a series of failed attempts to restore sexual function in rich, elderly men by transplanting slices of ape testes.

But it was Ivanov who made the most serious efforts at combining human and nonhuman apes. Earlier in his career, in addition to the successful artificial insemination of horses, Ivanov had created a variety of animal hybrids, including "zeedonks" (zebras + donkeys) and different combinations of small rodents (mice, rats, and guinea pigs). For a time in the 1990s a fictional version of Ivanov was the chief character in a Russian-era television show portraying him as the "Red Frankenstein."

In 1910, Ivanov had announced at a World Congress of Zoologists in Graz, Austria, that it might be possible to produce a human-ape hybrid via artificial insemination. During the mid-1920s, working at a laboratory in Conakry (then part of French Guinea) under the auspices of France's highly respected Pasteur Institute, Ivanov attempted just that, seeking without success to inseminate female chimpanzees with human sperm. (We don't know whose, and we also presume—although don't know for certain—that the attempted insemination was by artificial rather than natural means.) Then, in 1929, at the newly established Sukhumi Primate Research Institute, he endeavored to reverse donor and recipient, having obtained consent from five women volunteers to be inseminated—once again, presumably by artificial methods—with sperm from chimpanzees and orangutans. Inconveniently, however, the nonhuman primate donors died before making their "donations," and for reasons that are unclear, Ivanov himself fell out of political favor and was sent to Siberia in 1930; he died a few years later.

Ilya Ivanov's story is not especially well known outside Russia, and insofar as Westerners learn of it, they are inclined to ridicule it as an absurd, antique episode of reaching for a would-be "planet of the (communist) apes," or—paradoxically—to inveigh against the immorality of such at attempt, which is increasingly feasible. To be sure, Ivanov's crude efforts at cross-species hybridization are at present no closer to fruition, simply because even though human and chimp DNA are overwhelmingly similar, getting sperm from either species to combine with eggs from the other is—to put it literally—inconceivable. However, CRISPR makes it extremely likely that a humanzee could be generated in a laboratory. Such an individual would not be an exact equal-parts, 50-50 hybrid, but would be neither human nor chimp: rather, something in between.

A quick note on terminology. A hybrid is a cross between individuals of distinct genetic ancestry, which means that technically, nearly everyone is a hybrid, except for clones, identical twins, or perhaps persons produced by close incest. More usefully, we speak of hybridization as the process by which members of different subspecies are crossed (mating Labradors and poodles, for example, to produce labradoodles), or—more rarely—different species, in which case the resulting hybrids are often nonviable, either sterile (e.g., mules, hybrids made by crossing horses and donkeys), or just plain unusual (e.g., tigrons, which have occasionally been generated by hybridizing tigers and lions, or ligers, produced vice versa). In nearly all cases, hybrids are genetic mixtures, with essentially all body cells containing equal quantities of DNA from each parent. This, of course, is true of all sexually produced individuals: it's just that with identified hybrids, those two parents are likely to be more distantly related than is usual.

Chimeras, on the other hand, are somewhat different. They derive from what is essentially a process of grafting, whereby two genetic lines (most interestingly, different species) are combined to produce an individual that is partly of one genotype and partly of another, depending on which cells are sampled, and at what point in embryonic development. Probably because it is easier to imagine creatures produced by combining identifiable body parts from different animals than to picture a mingled, intermediate form, chimeras, more than hybrids, have long populated the human imagination. Ganesh, the Hindu god with a human body and elephant's head, is a chimera, as are the horse-human centaurs of Western mythology. The classic "chimera" of Greek legend had the head and body of a lion, a tail that had morphed into the head of a snake and—to make a weird creature even more so—the head of a goat, sometimes facing forward and sometimes backward.

It is unclear whether my own imagined chimphuman will be a hybrid (produced by cross-fertilizing human and nonhuman gametes à la Ivanov), or a chimera, created in a laboratory via techniques of genetic manipulation. I'm betting on the latter.

Even now, research efforts are underway seeking to produce organs (kidneys, livers, etc.) that develop within an animal's body—pigs are the preferred target species—and whose genetic fingerprints are sufficiently close to the *Homo sapiens* counterpart to be accepted by a human recipient's immune system while also able to function in lieu of the recipient's damaged organ. A human skin cell, for example, can be biochemically induced to become a "pluripotent stem cell," capable of differentiating into any human tissue type. If, say, a replacement liver is desired, these stem cells can be introduced into a pig embryo after first using CRISPR to inactivate the embryo's liver-producing genes. If all goes well, the resulting pig-human chimera will have the body of a pig, but containing an essentially human liver, which would then be available for transplantation into a person whose liver is failing.

After years of opposition, the US National Institutes of Health announced in August 2016 that it intends to lift its moratorium on such research, which holds out promise for treating (perhaps even curing) many serious human diseases, such as cirrhosis, diabetes, and Parkinson's. Currently prohibited—and likely to remain so—is federal funding for studies that involve injecting human stem cells into embryonic primates, although inserting such cells into adults is permissible.

My recommendation in this regard will be not only controversial but also, to many people, downright immoral: I applaud any and all biomedical research that involves the creation of human-nonhuman chimeras or hybrids, and not simply because of the potential disease-curing benefits this might offer. I submit that generating humanzees or chimphumans would be not only ethical, but profoundly so, even if there were no prospects of enhancing human welfare, because it would go a long way toward overcoming what may be the most hurtful, scientifically invalid and deeply immoral myth of all time, namely that human beings are qualitatively discontinuous from other living things.[a] It is hard to imagine how even the most determined homocentric, animal-denigrating religious fundamentalist could maintain that god created us in his image and that we and we alone harbor a spark of the divine once confronted with living beings that are indisputably intermediate between human and nonhuman.

No one knows precisely what motivated Ilya Ivanov's early fertilization experiments. Maybe it was the allure of the possible, such that having discovered the potent hybrid-generating hammer of in vitro fertilization, everything—including eggs and sperm, with one from human and the counterpart from nonhuman primates—looked alluringly like a nail. Or maybe he was driven by the prospect of currying favor with Stalin, or of fame (or infamy) had he succeeded, or perhaps as an ardent atheist Bolshevik, Ivanov was inspired by the prospect of disproving religious dogma.

In any event, the nonsensical insistence that human beings are uniquely created in god's image and endowed with a soul, whereas other living things are mere brutes has not only permitted but also encouraged an attitude toward the natural world in general and other animals in particular that has been at best indifferent and more often, downright paternalistic, antagonistic, and in many cases, intolerably cruel. It is only because

---

[a] Shortly after writing this, I discovered that Richard Dawkins had made a similar suggestion (https://www.theguardian.com/science/blog/2009/jan/02/richard-dawkins-chimpanzee-hybrid). I am delighted at this convergence.

of this self-serving myth that some people have been able to justify keeping other animals in such hideous conditions as factory farms in which they are literally unable to turn around, not to mention prevented from experiencing anything approaching a fulfilling life. It is only because of this self-serving myth that some people accord the embryos of *Homo sapiens* a special place as persons-in-waiting, magically endowed with a notable humanity that entitles them to special legal and moral consideration unavailable to our nonhuman kin. And it is only because of this self-serving myth that many people have been able to deny the screamingly evident evolutionary connectedness between themselves and other life forms.

When claims are made about the "right to life," invariably the referent is *human* life, a rigid distinction only possible because of the presumption that human life is somehow uniquely distinct from other forms of life, even though everything we know of biology demonstrates that this is simply untrue. What better, clearer, and more unambiguous way to demonstrate this than by creating viable organisms that are neither human nor animal but certifiably intermediate?

But what about those unfortunate individuals thereby produced? Neither fish nor fowl, wouldn't they find themselves intolerably unspecified, doomed to a living hell of biological and social indeterminacy? This is possible, but it is at least arguable that the ultimate benefit of teaching human beings their true nature would be worth the sacrifice paid by a few unfortunates. In addition, it is also arguable that such individuals wouldn't necessarily be so unfortunate at all. For every chimphuman or humanzee frustrated by her inability to write a poem or program a computer, there could equally be one delighted by her ability to do so while swinging from a tree branch. And—more important—for any human being currently insistent on his or her species specialness, to the ultimate detriment of literally millions of other individuals of millions of other species, such a development would be a real mind expander and paradigm buster.

Old paradigm: Human beings, presumably because they have divine souls, should never be genetically combined with other animals, who don't.

New paradigm (not really a paradigm so much as an impertinent suggestion): Creating a new and viable organism by combining human and nonhuman DNA might usefully open otherwise closed minds to the connectedness of human beings and other living things.

NOTE

1. Maxim Gorky. *Untimely Thoughts: Essays on Revolution, Culture, and The Bolsheviks, 1917–1918* (New Haven, CT: Yale University Press, 1995).

8

Separateness of Self?

WHEN IT COMES to mind expanders, here is another one, even larger than enhancing the boundaries of what it means to be human. At issue: what it means to be a living thing at all—that is, a "thing" that isn't really separate from everything else. It's a stretch, both mentally and literally.[1]

The issue is superficially simple, easily imagined but also easily misunderstood. Everyone knows that he or she—and every other living entity—is a distinct organism, separate from every other. An open-and-shut case, until you pay attention to what we have finally learned from basic physiology, ecology, and evolution (and what Buddhists have recognized for more than two thousand years): although each of us lives within his or her body, and in that trivial sense everyone is a distinct, skin-encapsulated ego, this sense of rigid individuality is nonetheless an illusion. For example, our insides and "the outside" are constantly being exchanged and transferred, nearly always without our knowing it. As the old ad for milk used to announce, "There's a new you coming every day," so what yesterday was "out there" is today "in here." And tomorrow?

The joke asks, "What did the Buddha say to the hot dog vendor?" Answer: "Make me one with everything." Condiments aside, the Buddha knew that he was already intimately connected to the rest of reality. The Sanskrit word is *anatman*, or "nonself," one of the most difficult Eastern concepts for Westerners to grasp. Easily caricatured, it can also be readily grasped at last, especially with the aid of modern, mostly Western science.

We and all other organisms aren't like apricots or cherries, with a hard, separate, inner core that is walled off from everything and everyone else. Instead, we are like onions. When our superficial "selves" are peeled away, there is no underlying inner nucleus of self-stuff, no tiny, fully contained, and self-sufficient homunculus, separated from the rest of the organic and inorganic world. This deep illusion-crashing reality can be seen not only in the process of respiration and metabolism, as well as in basic ecological interdependencies, but also in evolution itself, by which genes cascade in a flowing stream from our ancient ancestors to ourselves, en route to their next manifestation in subsequent, seemingly separate and independent "individuals." Think of a whirlpool: composed of what appears to be definite boundaries, it actually exists only as a temporary flow-through of matter and energy. That's what "each" of us really is.

More than forty years ago, in his book *The Lives of a Cell*, Lewis Thomas marveled at how intimately we are "embedded in nature." This image falls a bit short, however, since "embedded" implies a separate and distinct thing stuck in a discernibly different matrix, like a fossil fly encased in amber. Dr. Thomas nonetheless knew whereof he wrote, since he explained, "A good case can be made for our nonexistence as entities . . . . We are shared, rented, occupied" by "little separate creatures."[2] Even here, however, much as he was an admirable researcher, thinker, and writer, Lewis Thomas didn't go far enough. Those "separate creatures" are indeed little, but they aren't truly separate. Nor are we.

One needn't be a Buddhist or a scientist to recognize this, although it helps to access some of the insights of both. Here is Ralph Waldo Emerson, in his essay *Nature*: "Strictly speaking, . . . all that is separate from us, all which philosophy distinguishes as the NOT ME, that is, both nature and art, all other men and my own body, must be ranked under this name, NATURE." The science of ecology did not exist in Emerson's day, but it does today and it speaks loudly to anyone willing to listen.

THICH NHAT HANH is a beloved Vietnamese monk in the Zen tradition who was nominated for a Nobel Peace Prize by Martin Luther King Jr. and who has been strongly influenced by the wisdom of ecology along with that of Buddhism. Accordingly, he has proposed the word—and, more importantly, the concept—of "interbeing" as a fundamental insight. "If you are a poet," he writes,

> you will see clearly that there is a cloud floating in this sheet of paper. Without a cloud, there will be no rain; without rain, the trees cannot grow; and without trees, we cannot make paper. The cloud is essential for the paper to exist. If the cloud is not here, the sheet of paper cannot be here either. So we can say that the cloud and the paper inter-are. "Interbeing" is a word that is not in the dictionary yet, but if we combine the prefix "inter-" with the verb "to be," we have a new verb, inter-be. Without a cloud and the sheet of paper inter-are. If we look into this sheet of paper even more deeply, we can see the sunshine in it. If the sunshine is not there, the forest cannot grow. In fact, nothing can grow. Even we cannot grow without sunshine. And so, we know that the sunshine is also in this sheet of paper. The paper and the sunshine inter-are. And if we continue to look, we can see the logger who cut the tree and brought it to the mill to be transformed into paper. And we see the wheat. We know the logger cannot exist without his daily bread, and therefore the wheat that became his bread is also in this sheet of paper. And the logger's father and mother are in it too. When we look in this way, we see that without all of these things, this sheet of paper

cannot exist. Looking even more deeply, we can see we are in it too . . . So we can say that everything is in here with this sheet of paper.

You cannot point out one thing that is not here: time, space, the earth, the rain, the minerals in the soil, the sunshine, the cloud, the river, the heat. Everything co-exists with this sheet of paper . . . "To be" is to inter-be. You cannot just be by yourself alone. You have to inter-be with every other thing. This sheet of paper is, because everything else is. Suppose we try to return one of the elements to its source. Suppose we return the sunshine to the sun. Do you think that this sheet of paper will be possible? No, without sunshine nothing can be. And if we return the logger to his mother, then we have no sheet of paper either. The fact is that this sheet of paper is made up only of "non-paper elements." And if we return these non-paper elements to their sources, then there can be no paper at all. Without "non-paper elements," like mind, logger, sunshine and so on, there will be no paper. As thin as this sheet of paper is, it contains everything in the universe.[3]

This, then, is the key to seeing beyond the paradigm that insists on separateness of self. To recognize the new paradigm of interbeing is not to deny the obvious truth that we perceive ourselves as separate, independent beings, but to understand that this separateness is an illusion. It is a useful one, to be sure, enabling people to conceptualize how their own superficially distinct piece of paper (labeled, by convention, "me" or "I") navigates its many interconnections. It also enables biologists to perform their research, distinguishing an elk from a wolf, and separating each from, say, a beaver or a maple tree. And there is much to be learned about how a wolf goes about hunting elk, or how beavers go about constructing dams and building their lodges.

But wolves don't just depend on elk and other species as prey; they literally contain elk and their ilk within themselves, just as Thich Nhat Hanh's piece of paper contains the whole world. In 1926, the last wolf was killed in Yellowstone National Park, in what we now know was a foolish effort to "improve" the environment for—among other species—elk. As a result, the elk population soared, but lots of other things happened, too. The increased numbers of elk, no longer alert to possible predators, congregated especially along streams and rivers, where in the past they could be ambushed. There they overbrowsed trees such as cottonwoods, aspen, and willows, whose presence had restrained riverbank erosion. With these plants diminished, waterways became broader and therefore shallower, which as a result grew less shaded and warmer. The leaves of aspen in particular were consumed when just beginning to sprout and so the trees rarely attained their full height. Songbirds were accordingly deprived of nesting sites and their number diminished. Beavers, which depend on healthy trees as well as deep water, became increasingly rare until there was just one beaver colony left in the entire park. And trout, which require cold water (largely because higher temperature reduces water's ability to hold dissolved oxygen), began to decline in those wider, shallower, warmer rivers and streams.

Meanwhile, coyote numbers increased, because wolves were no longer competing with (and eating) them. Those abundant coyotes, although too small to have a significant effect on the exploding elk population, chowed down on small mammals such as ground squirrels, voles, mice, and pocket gophers, whose populations crashed, bringing with them the numbers of foxes, badgers, owls, hawks, and falcons, all of whom depended on an abundance of small mammals in order to prosper.

Then, beginning in 1995, wolves were reintroduced into the park, and its entire life-fabric changed. Park rangers had been forced to cull (i.e., kill) increasing numbers of elk, but now the wolves took over this task. Not only did the elk population decline, approaching its pre-wolf-holocaust numbers, but their behavior reverted as well: they became more vigilant, especially around watercourses, also spending more time on knolls and ridges where they could spot predators. The severely overgrazed streamside vegetation quickly recovered, and the water became deeper, cooler, and healthier for its inhabitants. Aquatic insects such as mayflies and stoneflies returned, and with them, the trout population. Wildflowers came back in great abundance, and in their wake, an array of terrestrial and airborne insects, on which songbirds began feasting once again, just as they returned to nesting in the larger, healthier trees.

Coyote numbers declined, since they were not only outcompeted by the reintroduced wolves, but also, on occasion, hunted by them. This was a tribulation for the coyotes, but a benefit to pronghorns, which were always too fast to be threatened by wolves, but whose calves were hunted by coyotes. Not only that, but with fewer coyotes—and because the wolves were concentrating on larger prey such as deer and elk—the population of once abundant small mammals (those ground squirrels, mice, voles, and so forth, who were a preferred food of coyotes), returned to its previous levels, enabling foxes, badgers, and raptorial birds to make a living once again.[4]

One way of looking at this, which is only a small subset of all the interactions occurring in the greater Yellowstone ecosystem following the reintroduction of wolves, is that wolves, elk, coyotes, beaver, fish, birds, insects, trees, rivers, and so forth are mutually interacting and interdependent, such that impacting one component sets off reverberations in all the others. The English poet Francis Thompson put it this way in his poem "The Mistress of Vision":

All things . . . near and far
hiddenly to each other connected are,
that one cannot touch a flower without the troubling of a star.

Another and more biologically accurate way is to expand our perceptions and acknowledge that the reason you cannot touch a flower without troubling a star, or exterminate wolves without troubling, for example, badgers,[a] is that there literally is no meaningful distinction between the former and the latter. The "fact" that individuals are separate seems so obvious as to allow no refutation, but bear in mind that many seemingly obvious natural facts are not as they appear: the Earth looks flat, but it isn't. The sun appears to move around the Earth, but it doesn't. That piece of paper on which you might be reading this book (or the Kindle or other device) certainly looks solid, but physics tells us with equal certainty that it is made almost entirely of empty space, with a comparatively infinitesimal amount of solid stuff sprinkled here and—mostly—there.

Individual plants and animals may seem to be in fact, individuals, but on deeper reflection, they are revealed to be part of a larger, almost unlimited whole, whose "individuality" is more a matter of convenience and cognitive deception than meaningful reality. More troublesome is that insofar as we fail to recognize our connectedness, we risk doing harm—not only to ourselves but also, unavoidably, to the entire fabric on which all depend. "It really boils down to this," according to Martin Luther King Jr.: "All life is interrelated. We are all caught in an inescapable network of mutuality, tied into a single garment of destiny. Whatever affects one destiny, affects all indirectly."[5]

Most people nonetheless find it easier (more convenient, perhaps, to their psychic comfort) to accept a lack of boundaries between animals and their environment than when considering human creatures. But people are as immersed in this boundarylessness as are all other beings. *Homo sapiens* occupying a modern city might appear less connected to each other and to their surroundings than are wolves occupying the Lamar Valley in Yellowstone National Park, but they are no more disconnected than were our Australopithecine ancestors occupying a Pleistocene-era African savannah. And although our wolf-ecosystem example focused on food, this was simply because the eat-and-being-eaten linkage is more immediately apparent and thus easier to, well, digest than are the myriad other interpenetrations. Food webs constitute just one case in which distinctions between self and not-self melt away on deeper examination. Moreover, this array of interconnectedness isn't only true of ecological networks and linkages.

One of the most important advances in modern evolutionary biology is the recognition that natural selection proceeds not by relative success and failure of individual bodies but of their constituent genes. Whether conceptualized as selfish or altruistic,

---

[a] Badgers rely, among other things, on ground squirrels, which had become rare because of the increase in coyotes, which themselves had increased in abundance because of the elimination of wolves.

there is no doubt that these genes ultimately create bodies and not vice versa. Bodies—those structures that masquerade as distinct individuals—have been brilliantly characterized by Richard Dawkins as survival machines generated by genes for their (the genes') perpetuation. It is genes, not bodies, that persist over evolutionary time, since identifiable organisms are necessarily short-lived, existing only in a brief slice of time as temporary storage sites and ultimately as evolutionary sling-shots, whose role is to project copies of genes into other, future bodies, which themselves will only persist for a comparatively brief time.

"We are but whirlpools in a river of ever-flowing water," wrote Norbert Wiener, the founder of cybernetics and one of the most important mathematicians of the twentieth century. Wiener went on: "We are not stuff that abides, but patterns that perpetuate themselves."[6]

The boundaries of these patterns are themselves indistinct. There is no reason, for example, to insist that they correspond to the easily identified boundaries of each individual body. Thus, we grow hair, which is clearly part of each "us." This hair, in turn, can be cut. But just because a haircut doesn't hurt doesn't mean that your hair has lost its connectedness to your "self." When it lies on the barbershop floor, it may no longer appear to be part of you, but it carries the imprint of your DNA, plus a record of the chemicals you have ingested or otherwise experienced.

When a lock of hair is then thrown "away," at what point does it cease being part of us? How about when some of its atoms are incorporated into an earthworm or gull scavenging at the local landfill? When a spider weaves a web, which seems separate from the spider's body, isn't her web as much a consequence of that spider's complexly interacting genotype as is the animal's body? Where, then, is the line separating spider from web, or web from branch to which it is attached, and so on? And where, similarly, is the line separating each human "being" from, say, a squirt of that being's urine, or blob of feces, or for that matter, from the human equivalent of a spider's web or a beaver's dam—say, a log cabin, a city or a spaceship, none of which could be constructed without the intimate involvement of human DNA, and which are literally extensions of the body that created them, just as a beaver dam is an extension of a beaver.

When Dawkins wrote his book *The Extended Phenotype*, his goal was to emphasize the long reach of genes, which goes beyond involvement in merely generating a body, to include those nonbody items that wouldn't exist without genes functioning somewhere in the background. At the same time, however, he was illustrating yet another dimension in which genes as well as bodies extend into the rest of the world.

Don't be misled about those genes, either. No one swims outside the gene pool. Moreover, these genes are inseparable from their environments, which includes other genes affiliated on the same chromosome, not to mention the regulatory impact of additional stretches of DNA, RNA, proteins, histones, and an array of interacting chemical entities whose function—and even, whose existence—is only now being revealed. Not only that, but every gene is composed of numerous atoms, of which hydrogen is the most abundant. And hydrogen ions, we now know, undergo a regular and incredibly rapid-fire process of electron exchange, whereby a given entity labeled "hydrogen" is continually sharing electrons with its neighbors. As a result, not only bodies but also genes themselves are literally transient, both because they are subject to unpredictable copying errors (mutations) and because even temporarily stable chunks of DNA are always in flux, now you see them now you don't—now they are one thing but almost immediately afterward, they have swapped a crucial part of them-"selves" with whatever is next door. But in fact, there is no wall and thus, no "door."

So much for the supposed solidity and individuality of genes. Even though it is easy to grant the interconnectedness of populations such as wolves, elk, and aspen, it is equally tempting to draw the line at bodies themselves. After all, it is bodies—not genes or predator-prey populations—that we see out there in the macroscopic world. And each of those bodies seems indubitably bounded by its own private and personal partition of fur, feathers, scales, bark, or skin. As a song from the 1960s by the parodist Alan Sherman proclaimed, "You Gotta Have Skin"—to the tune of an even older song, "You Gotta Have Heart":

> You gotta have skin.
> All you really need is skin.
> Skin's the thing that if you've got it outside,
> It helps keep your insides in.
> You gotta have skin.

Here we see a comical but nonetheless accurate representation of the older paradigm, which argues for individuality, not merely to keep our insides in but as a simple matter of logic. But logic can be flawed, as it is in this case. Each person's skin is itself permeable, studded with holes (pores) through which gases and liquids are constantly entering and leaving. It isn't an exaggeration to note that a person's skin doesn't separate her *from* so much as join her *to* her surroundings. Imagine a sculptor who makes a marble statue of a horse by chipping away everything in a given block of stone that isn't a horse. The thing-in-itself is already present, and simply needs to be revealed.[b] There is also no hard and fast rule that commands precisely where in the marble resides the horse's left front hoof. It is distributed throughout; just as we are.

[b] Is it going too far to suggest that other creatures, an almost infinite array, are also simultaneously present within that block of stone, making it arbitrary what a particular artisan chooses to reveal? (Maybe so, but it's fun nonetheless!)

By the same token, few things appear as solid and distinct as a tree. Rewriting Joyce Kilmer, "I think that we will never see / an object separate as a tree." And yet, even those big, beautiful aspens whose maturation is attributable (at least in part) to a reduction in grazing by overcrowded, desperately hungry, wolf-deprived elk—even these trees turn out to be so connected to each other, via their anastomosing root systems that nearly every aspen grove can accurately be described as a unified aspen-organism rather than composed of separate, individual trees. Of course, it is possible to cut down individual trees, just as individual human lives are generated and, in time, cut down.

Just as supposedly individual bodies are less individual and more cross-linked than meets the eye, there is also a physical, cross-generational continuity between parents and offspring, be they trees connecting via pollen, flower, and seed, or people connected to their ancestors and descendants by sperm and egg. It isn't surprising that abortion occasions such conflict and confusion, not just theologically and politically but also logically and biologically, since there is no clear point at which a new individual not only becomes alive but also warrants being considered an individual. In the United States, legal precedent dictates that a developing fetus is entitled to a degree of personhood when he or she can survive independent of its mother, but in fact, no one is ever independent—of their mothers, fathers, earlier ancestors, the farmer who produces her food, the rain forest plants that generate atmospheric oxygen, all those fellow inhabitants of Thich Nhat Hanh's vast interbeing sheet of cosmic paper. Or, as Boris Pasternak observes in his poem, "Steppe," "You cannot go over the road past the fence / Without trampling the universe."[7]

To REPEAT: THE scientific recognition that individuals are merely a convenient fiction maps directly to the Buddhist conception of *anatman* or non-self. As I wrote in my book *Buddhist Biology*, nonself is not the same as nonexistence. Buddhists aren't fools. "If we arrive at the knowledge that the self at which we grasp is empty," writes the Dalai Lama,

we may imagine this means that we as individuals with personal identities do not exist. But of course this is not the case—our own personal experiences demonstrate that as subjects and agents of our own lives, we certainly exist. So how, then, do we understand the content of this insight into absence of self? What follows from this insight? We must be very clear that *only the self that is being grasped as intrinsically real* needs to be negated. The self as a conventional phenomenon is not rejected. This is a crucial aspect of the Buddha's teachings on emptiness. Without understanding this distinction, one cannot fully understand the meaning of no-self.[8]

It may seem peculiar, even psychotic, to claim that something as obvious as one's self isn't "intrinsically real." Remember, however, the illusory reality of a piece of paper, which masquerades as something closed off unto itself and yet is composed—as Thich Nhat Hanh puts it—entirely of "non-self elements." Or the illusion that a single elk, wolf, aspen, or person can be isolated in any meaningful sense from its, his, or her surroundings.

Moreover, we often encounter "things" that are ostensibly real but actually lack any intrinsic reality. A country, for example, such as Mexico, the United States, or Canada, is typically treated as though it exists—and when it comes to various markers of national identity such as borders, political organization, or currency, that existence is undeniable. But we all know that these distinctions are artificial impositions in which a horse has been arbitrarily isolated from its surrounding marble matrix and labeled "Mexico," "the United States," or "Canada," for human convenience, in many ways imposing a phony separateness on an underlying unity.

Donald Trump may bloviate about building a wall between the United States and Mexico, but, as Robert Frost pointed out, "Something there is that doesn't love a wall," not least because in addition to being temporary and in the long run counterproductive, walls represent an ultimately failing attempt to impose conventional isolation on underlying, indivisible unity. Distinctions that are imposed and often destructive become reified and made to seem "more real" than they actually are. Something very much like this has happened with respect to the standard Western conception of what it means to be a separate and distinct "self."

As I noted, however, in *Buddhist Biology*,

> Western thought has not been monochromatic when it comes to selfhood and individuality. Friedrich Nietzsche called for *Selbstuberwindung*, or "self-overcoming," just as the Buddhist *Dhammapada* teaches that "although he may conquer a thousand times a thousand men in battle, yet he indeed is the noblest victor who would conquer himself," and which in turn is paralleled by Islam's distinction between the "lesser" and the "greater" jihad, with the former defined by war against Islam's external enemies and the former, the more difficult struggle with one's self.[9]

Or as Goethe wrote in his poem, "The Mysteries":

> All force strives forward to work far and wide
> To live and grow and ever to expand;
> Yet we are checked and thwarted on each side
> By this world's flux and swept along like sand;

In this internal storm and outward tide
We hear a promise, hard to understand:
From the compulsion that all creatures bind,
Who overcomes himself, his freedom finds.[10]

It isn't always easy to overcome one's "self," which shouldn't be surprising, insofar as doing so actually requires struggling with unreality, trying to overcome a shadow. Sometimes, it is downright impossible. David Hume expressed one aspect of this:

> When I enter most intimately into what I call *myself*, I always stumble on some particular perception or other, of heat or cold, light or shade, love or hatred, pain or pleasure. I can never catch *myself* at any time without a perception, and never can observe any thing but the perception.

Despite the universality of this experience, most people spend their lives acting and thinking as though some kind of isolated internal self nonetheless exists, a solitary audience of one, a central intelligence agency inside each person's head, privately observing the constant play of perceptions and sensations. There is, instead, a connected and temporary aggregation of memories, cohering in space via elaborate neural networks—entirely composed of nonneural components—and misleadingly coherent in time as well, even though these neurons are constantly being weeded out, and their synaptic connections continually rearranged.

Charles Sherrington was a pioneering neurobiologist who discovered and named the synapse, the space between neurons through which various neurotransmitters pass and which thus connects them physiologically. He was deeply aware of the artifice involved in claiming that the aggregate of nerve cells that we label a "brain" is meaningfully equated with a separate and isolated self:

> This self . . . regards itself as one, others treat it as one. It is addressed as one, by a name to which it answers. The Law and the State schedule it as one. They identify it with a body which is considered by it and them to belong to it integrally. In short, unchallenged and untargeted conviction assumes it to be one. The logic of grammar endorses this by a pronoun in the singular. All its diversity is merged in oneness.[11]

Making things even more complex—or bizarre, or intriguing, or wonderful, depending on point of view—is the fact that Sherrington's (and the Buddha's) "oneness" isn't only irrevocably and seamlessly connected to the greater oneness out there in the world, but actually isn't a unitary "it" at all. Biologists have thus found, to their amazement, that each human being is host to literally trillions of other cells, many of them constituting the intestinal microbiome, a vast assemblage of cells that are genuinely "nonself" in that their genetic make-up is discriminable from the genotype of any human being—but without which no "one" could survive.

Even cells themselves—at least those associated with what we immodestly label "advanced organisms" such as us—that are designated "eukaryotic" (possessing a nucleus) are now widely recognized to have originated in a long-ago process whereby early ancestral cells became what they are today by literally engulfing other, "primitive," free-living organelles. This idea, designated "endosymbiosis theory," was originated by the late biologist Lynn Margulis[12] (who was also Carl Sagan's first wife). It holds that mitochondria, for example, had existed as early bacteria, which were literally consumed by other cells. The resulting, well-fed cells thereby obtained the energy-generating benefits provided by mitochondria, which are currently the power-plants of eukaryotic cells. A similar process seems likely to have involved the consumption of structures that, once incorporated into early plants, became chloroplasts and enabled photosynthesis.

Mitochondria and chloroplasts currently have DNA that is quite different from the "normal" DNA found in a cell's nucleus. Among higher animals, this mitochondrial DNA doesn't combine at fertilization with any contribution from the father, because sperm only provide nuclear DNA, and essentially none of the cellular matrix that is contained in the egg, all of which is inherited only from the mother. This mitochondrial DNA is thus inherited—unchanged except for occasional mutations—in a continuous maternal line. Over time, what had been aerobically metabolizing protocells became an essential part of those cells that constitute us today. The take-home message to be grasped by those neurons constituting the brains of those readers of this book whose neural energy—along with muscular, digestive, immunologic, and so forth—is provided by such "endosymbionts," which are now crucially incorporated into all human body cells is simply this: you are crucially composed of other cellular structures, which are now symbiotically cooperating to keep "each" of you going.

This "going" is altogether dependent on metabolic processes that use energy, which is provided by the process of metabolism, the constant breakdown of material whose energy is released in those mitochondria and converted into a simple molecule (called ATP) that serves as a kind of universal energy currency. This process, in turn, is necessitated by the Second Law of Thermodynamics, which recognizes that entropy (disorder) constantly increases in a closed system. Accordingly, life—which is a complex, nonrandom kind of structure—cannot be closed. Matter—namely, food—and the ability to access the energy contained therein requires a continuing through-put of not only food but also the molecules (usually oxygen) needed to "burn" that food. The wage of inaction, of stasis, of nonparticipation in the flux essential to life, is death.

The above makes mitochondria doubly fascinating when it comes to emphasizing the interconnectedness of life: on the one hand, they are the energy factories needed to keep each open-ended living apparatus functioning, while on the other, their very existence embodies a long history of open-endedness and permeable engulfment.

We, along with all living things, have unintentionally and unavoidably bought into a kind of time-share arrangement, in which we all take turns incorporating and then giving up the basic building blocks of matter itself. The astronomer Neil de Grasse Tyson[13] makes the point, for example, that there are roughly as many molecules of water in a cup as there are cups of water in all the world's oceans. This means that given enough time and mixing, it is reasonable to suppose that a molecule of water inside a drop of sweat beading your forehead had at one time coursed through the kidneys of Moses, Socrates, Jesus, and/or Mohammed, not to mention a *Tyrannosaurus rex*. By the same token, since there are roughly the same number of air molecules in a puff of human breath as there are breaths in the Earth's atmosphere, some of the air you breathe in and out would likely have been similarly respired by a brontosaurus and/or a woolly mammoth.

At this point, it seems fair to ask "So what"? What are we to make of the fact that as biologists have recently confirmed and as John Donne observed poetically four centuries ago, "No man is an island, entire unto himself"? It is one thing to accept, albeit begrudgingly, that we are smaller and less central to the cosmos than previously thought (Tyson calls it a "cosmic perspective"), that our history unquestionably connects us to the rest of the living world, and, moreover, that the universe doesn't really seem to exist with us in mind. But it is yet another step to acknowledge that even the edges and boundaries of our own proudly private and personal existence—right here and now, as you read this—are not only indistinct but literally nonexistent. The prospect is dizzying, threatening to leave us, once again as Nietzsche protested, unanchored and without reliable guidelines or compass points.

One option is to shrivel under the stern gaze of our own anthropo-diminution. Another is to ride the wave of connectedness to an enhanced sense of our responsibility to each other and to the rest of the world, since we are, as no less a scientist than Albert Einstein noted, "part of a whole, called by us the 'Universe'—a part limited in time and space. [Each person] experiences himself, his thoughts, and feelings, as something separated from the rest—a kind of optical delusion of his consciousness."

Einstein went on to observe,

> This delusion is a kind of prison for us, restricting us to our personal desires and to affection for a few persons nearest us. Our task must be to free ourselves from this prison by widening our circles of compassion to embrace all living creatures and the whole of nature in its beauty.[14]

One needn't be a Buddhist master, a biologist, or a pioneering physicist to feel a responsibility to embrace all living creatures and the whole of nature. Moreover, this responsibility needn't be embraced with a sense of dutiful reluctance; rather, as Einstein proposed, it lends itself equally to an appreciation of its beauty. It is an undertaking worthy of poetic joy.

Not coincidentally, the greatest American poet, Walt Whitman, contained all of these instincts, to which he added a heady dose of effervescent egotism, notably in his masterwork, *The Song of Myself*, which begins—true to its title—as follows: "I celebrate myself, and sing myself / and what I assume you shall assume." Whitman then segues to the following crucial recognition, worthy of a biologist, physicist and Buddhist: "For every atom belonging to me as good belongs to you." Later, he celebrates the wonder and beauty of the interconnected natural world:

> I believe a leaf of grass is no less than the journeywork of the stars,
> And the pismire is equally perfect, and a grain of sand, and the egg of
>   the wren,
> And the tree toad is a chef-d'oeuvre for the highest,
> And the running blackberry would adorn the parlors of heaven,
> And the narrowest hinge in my hand puts to scorn all machinery,
> And the cow crunching with depress'd head surpasses any statue,
> And a mouse is miracle enough to stagger sextillions of infidels.

Whitman then joyously acknowledges his own debt to the stuff of shared reality, with more than a passing nod to evolutionary unity as well:

> I find I incorporate gneiss, coal, long-threaded moss, fruits, grains,
>   esculent roots,
> And am stucco'd with quadrupeds and birds all over.

Finally, in perhaps the literary world's best, briefest, and deepest statement of how self merges into nonself, and thus without boundaries and without regret into everything (but assuredly not nothing), we read:

> I bequeath myself to the dirt to grow from the grass I love,
> If you want me again look for me under your boot soles.

Old paradigm: Everyone is separate and distinct, an army of one.

New paradigm: Not so! The boundaries between individuals are arbitrary, artificial, and for the most part illusory. Our states are united.

NOTES

1. Some material in this chapter was revised and repurposed from my book *Buddhist Biology: Ancient Eastern Wisdom Meets Modern Western Science* (New York: Oxford University Press, 2014).
2. Lewis Thomas, *The Lives of a Cell* (New York: Viking, 1975).
3. Thich Nhat Hanh, *The Heart of Understanding* (Berkeley, CA: Parallax Press, 1987).
4. W. J. Ripple and R. L. Beschta, "Trophic Cascades in Yellowstone: The First 15 Years after Wolf Reintroduction," *Biological Conservation* 145, no. 1 (2012): 205–213.
5. Martin Luther King Jr., *The Essential Martin Luther King, Jr.* (Boston: Beacon Press, 2013).
6. Norbert Wiener, *The Human Use of Human Beings: Cybernetics and Society* (Boston: Houghton Mifflin, 1950).
7. From *Selected Poems of Boris Pasternak*, trans. Peter France (London: Allen Lane, 1983).
8. Tenzin Gyatsu (the Dalai Lama), *Essence of the Heart Sutra* (Boston: Wisdom Publications, 2005), italics in original.
9. D. P. Barash, *Buddhist Biology* (New York: Oxford University Press, 2014).
10. Johann Wolfgang von Goethe, "Mysteries," quoted in Walter Kaufmann, *Nietzsche: Philosopher, Psychologist, Antichrist* (Princeton, NJ: Princeton University Press, 1974).
11. Charles S. Sherrington, *The Integrative Action of the Nervous System* (Cambridge: Cambridge University Press, 1947).
12. Lynn Sagan (Margulis), "On the Origin of Mitosing Cells," *Journal of Theoretical Biology* 14, no. 3 (1967): 225–274.
13. Neil de Grasse Tyson, *Astrophysics for People in a Hurry* (New York: Norton, 2017).
14. Albert Einstein, as quoted in the *New York Times*, March 29, 1972.

Part II
New Ways of
Understanding
Human Nature

Prelude to Part II

THERE IS A marvelous scene toward the end of Proust's *In Search of Lost Time* in which Marcel, the protagonist, encounters people he hasn't seen in many decades, whereupon he marvels at how much older *everyone else* has become! It is difficult to look at ourselves honestly and without illusion, although it can also be bracing. Part 1 considered how anthropocentrism has been and continues to be challenged. Part II looks more narrowly at human nature and how it has been shaped by evolution, thereby reexamining an array of biology-based human nature paradigms that have been widely accepted but typically unquestioned, and which have been overthrown or are in the process of crumbling. The effect, for some, is to take our species down a peg or two by revealing the animals that we are; for others, who perceive that connection to a larger whole makes us not smaller but bigger, the effect is likely to be encouraging, even exhilarating.

One might expect that anthropologists, as the most biologically informed students of human behavior, would be especially predisposed to identifying and researching the evolutionary correlates of what people do, and why. Obviously "biofriendly" anthropologists, however, are generally paleontologists and physical anthropologists, suitably invested in studying the fossil record of primate and human evolution. Among most cultural anthropologists, by contrast, there is a persistent bias *against* cross-cultural universals, traits that—like bipedalism, opposable thumbs, an enlarged frontal cranium, and so forth—characterize *Homo sapiens* generally, regardless of social group, linguistic affinity, or geographic location. With a few notable exceptions, cultural anthropologists are more inclined to study, say, the details of human marriage systems than to note the fact that all human societies engage in some form of formalized bonding between adults.

It is indeed fascinating to note the specifics of how different societies engage in their own ceremonial procedures; moreover, such observations often constitute the take-home data obtained after lengthy and grueling fieldwork. But such a granular analysis also focuses attention away from those shared aspects of the human behavioral heritage that speak most loudly to our common evolutionary past—and therefore to a biological nature that we share with other species.

A basic grounding in evolutionary biology certainly helps an observer to appreciate the existence and significance of such panhuman traits. It also mitigates against the misleading claim that a biological perspective on human behavior somehow predisposes to racism, since if anything, a potent antidote to racism resides in identifying those behaviors that all *Homo sapiens* share, regardless of superficial differences in language, skin color, hair texture, eye shape, and so forth. Whatever the distinctions between people, they are undergirded by a solid foundation of shared, biologically generated human nature.

In his *Enquiry Concerning the Human,* and without benefit of Darwin, David Hume argued for the existence of such a solid, cross-cultural bedrock, using as examples the inhabitants of England and France (which, given the warlike competition between these two nations during the eighteenth century, must have seemed as far apart as two human groups could be). "Would you know the sentiments, inclinations, and core of life of the French and English," wrote Hume,

> you cannot be much mistaken in transferring to the former most of the observations which you have made with regard to the latter. Mankind are so much the same in all times and places that history informs us of nothing new or strange in this particular. Its [history's] chief use is only to discover the constant and universal principles of human nature, by showing men in all varieties of circumstances and situations, and furnishing us with materials from which we may . . . become acquainted with the regular springs of human action and behavior.

Part I of this book offered a wide-angle perspective, attempting to show that we really aren't the center of the universe (literally or figuratively). The common pattern in Part II is that although there is plenty to admire in human nature, all this admirable stuff is altogether *natural.* Rather than being qualitatively unique, what Hume called the "regular springs of human action and behavior" are woven from the same biological cloth as other living things.

The Buddhist monk and scholar Mathieu Ricard, who was initially trained as a molecular biologist, expressed it well, adding that with the death of special creation and anthropocentrism should also come the demise of anthropo-domination:

> Starting with the era of the ancestors we share with other animal species, little by little, by a long series of steps and minimal changes, we arrived at the stage of *Homo sapiens.* In the course of this slow evolution there was no "magical moment" that would justify our conferring on ourselves a special nature that makes us fundamentally different from the many species of hominids that

preceded us. Nothing occurred in the evolutionary process that would justify our claim to a right of total supremacy over the animals.[1]

Although included in his book *A Plea for the Animals*, Ricard's insight could as well have been titled "A Plea for the People," especially if accompanied by recognition that our connectedness to other life forms expands human boundaries rather than narrowing them. A similar view was endorsed by Pope Francis, who declared in his encyclical on the environment, "We must forcefully reject the notion that our being created in God's image and given dominion over the earth justifies absolute domination over other creatures. The Bible has no place for a tyrannical anthropocentrism."

Although this perspective is one that most scientifically literate people would endorse (even as most religious conservatives respond otherwise), it can lead in some uncomfortable directions, notably when traditional assumptions about the nature of human nature are thereby challenged. Thus, a widespread phenomenon—returning here to Brahe's blunder—is to accept the fact of human evolution, but to limit this insight to fossilized bones and archeological artifacts, as well as certain undeniable aspects of our embryology, anatomy, and physiology, while at the same time denying its relevance to what is paradoxically a "deeper" but more troubling aspect of our self-image: our own behavior.

AT THIS POINT, some humility is needed. Science differs from theology and the humanities in that it is made to be improved on and corrected over time. Hence, the new paradigms that follow are approximations at best, not rigid formulations of Truth and Reality; they are open to revision—indeed, much of their value resides precisely in their openness to modification in the light of additional evidence. By contrast, few theologians, or fundamentalist religious believers of any stripe, are likely to look kindly on "revelations" that suggest corrections to their sacred texts. In 1768, Baron d'Holbach, a major figure in the French Enlightenment, had great fun with this. In his satire *Portable Theology* (written under the pen-name Abbé Bernier, to hide from the censors), d'Holbach defined Religious Doctrine as "what *every good Christian must believe or else be burned*, be it in this world or the next. The dogmas of the Christian religion are immutable decrees of God, who cannot change His mind except when the Church does." Fundamentalist doctrines of Judaism and Islam are no different.

Such rigidity is not, however, limited to religion. For a purely secular case, most people today agree that it would be absurd to improve on Shakespeare, as was attempted, for example, by Nahum Tate, who rewrote the ending of *King Lear* in 1682, a "correction" that was approved by none other than Samuel Johnson, who agreed with Mr. Tate that the death of Cordelia was simply unbearable.

By contrast, science not only is open to improvements and modifications but also is to a large extent *defined* by this openness. Whereas religious practitioners who deviate from their traditions are liable to be derogated—and sometimes killed—for their apostasy, and even among secularists rewriting Shakespeare or playing Bach on an electric guitar is likely to be treated as indefensible, science thrives on correction and adjustments, aiming not to enshrine received wisdom and tradition but to move its insights ever closer to correspondence with reality as found in the natural world. This is not to claim that scientists are less vulnerable to peer pressure and the expectations of intellectual conformity than are others. We are, all of us, only human. But an important part of the peer pressure experienced by scientists involves openness to revision and reformulation. The Nobel Prize–winning ethologist Konrad Lorenz once wrote that every scientist should discard at least one cherished notion every day before breakfast. Although I don't recall Dr. Lorenz conspicuously following his own advice, it nonetheless captures a worthwhile aspiration.

Well-established scientific concepts aren't discarded lightly. Nor should they be. As Carl Sagan emphasized, extraordinary claims require extraordinary evidence, and so, it is reasonable that any new findings that go against received scientific wisdom—especially in proportion as those claims are dramatic and paradigm-shifting—should receive special attention, which includes being held to a high standard. But closed-mindedness is among the worst sins of an enterprise devoted to seeking truth rather than to validating old texts and anointed wisdom (religion) or passing along something that has been woven out of whole cloth (the creative arts).

In this book, I employ the word "truth" without quotation marks, because despite the protestations of some postmodernists and outright data-deniers, the natural world is undeniably real, and it is the noble endeavor of science to describe and understand this reality, which leads to the following proclamation—which shouldn't be controversial, but has become so, at least in some quarters: science is our best, perhaps our only way of comprehending the natural world, including ourselves. Moreover, we are approaching an ever more accurate perception of that world and of our place in it.

This argument, by the way, is not intended to glorify science or scientists at the expense of the creative arts and its practitioners. In fact, a case can be made that an act of artistic creativity is actually more valuable than a scientific discovery, not least because humanistic creativity yields results that, in the absence of their creators, are unlikely to have otherwise seen the light of day. If Shakespeare had not lived, for example, it is almost certain that we would not have *Hamlet*, *Othello*, *King Lear*, *Macbeth*, the hilarious comedies, those magnificent sonnets, and so forth. Without Bach, no *Goldberg Variations* or *Well-Tempered Clavier*, and none of his toccatas and fugues. No Leonardo? No *Mona Lisa*. The list goes on, and is as extensive as human creativity itself.

By contrast, although counterfactual history is necessarily speculative, an intuitive case can be made that if there had never been an Isaac Newton, someone else—perhaps his rival Robert Boyle or the Dutch polymath Christiaan Huygens—would have come up with the laws of gravity and of motion, which, unlike the blank pages on which Hamlet was written, or Leonardo's empty easel, were out there in the world, waiting in a sense to be discovered. Gottfried Leibniz, after all, has a strong claim to being at least the codiscoverer with Newton of calculus, which also was, equally, a truth waiting to be found. Unlike a Shakespeare or Leonardo, who literally created things that had not existed before and that owe their existence entirely to the imaginative genius of their creators, Newton's contributions were *discoveries*, scientific insights based on preexisting realities of the world (e.g., the laws of motion) or logical constructs (e.g., calculus).

Similarly, if there had been no Darwin, we can be confident that someone else would have come up with evolution by natural selection. Indeed, someone did: Alfred Russell Wallace, independently and at about the same time. A comparable argument can be made for every scientific discovery, insofar as they were—and will be—true discoveries, irrevocably keyed to the world as it is and thus, one might say, lingering just off-stage for someone to reveal them, like the "New World" waiting for Columbus. If not Columbus, then some other (equally rapacious and glory-seeking) European would have sailed due west from the Old World, just as if not Copernicus then someone else—maybe Kepler or Galileo—would have perceived that the solar system is solar- and not Earth-centric.

Mendelian genetics, interestingly, was rediscovered in 1900 by Hugo DeVries, Carl Correns, and Erich von Tschermak, three botanists working independently, more than a generation after Gregor Mendel published his little-noticed papers outlining the basic laws of inheritance. For us, the key point isn't so much the near simultaneity of this scientific awakening, but the fact that Mendel based his own findings on something genuine and preexisting (genes and chromosomes, even though they weren't identified as such until a half-century later), which, once again, were populating the natural world and waiting to be uncovered by someone with the necessary passion, patience, and insight.

Ditto for every scientific discovery, as contrasted to every imaginative creation. No matter how brilliant the former or the latter, practitioners of both enterprises operate under fundamentally different circumstances. It does not disparage Einstein's intellect to note that in his absence, there were many brilliant physicists (Mach, Maxwell, Planck, Bohr, Heisenberg, etc.) who might well have seen the reality of relativity, because no matter how abstruse and even unreal it may appear to the uninitiated, relativity is, after all, based on reality and not entirely the product of a human brain, no matter how brainy.

In some cases, the distinction is less clear between products of the creative imagination and those of equally creative scientific researchers. Take Freud, for example. His major scientific contribution—identifying the unconscious—is undoubtedly a genuine discovery; that is, the elucidation of something that actually exists, and had Freud never existed, someone else would have found and initiated study of those patterns of human mental activity that lie beneath or otherwise distinct from our conscious minds. On the other hand, it isn't at all obvious that absent Freud, someone else would have dreamed up infantile sexuality, the Oedipus complex, penis envy, and so forth. And this is precisely my point: insofar as alleged discoveries meld into imaginative creativity, they veer away from science and into something else.

Among some historians, there has been a tendency to assume that history proceeds toward a fixed point so that the past can be seen (only in retrospect, of course) as aiming at a superior or in some sense a more "valid" outcome. Such a perspective has been called "Whig history," and it is generally out of favor among contemporary scholars, as it should be. A notable example is the announcement by Francis Fukuyama that with the disbanding of the Soviet Union in 1991, we had reached the "end of history" and the final triumph of capitalist democracy.[2] Subsequent events have made the inadequacy of this claim painfully obvious. Despite the importance of democracy and its widespread appeal (at least in the West), it is important to be wary of the seductiveness of such "Whiggery," and to appreciate that changes in sociopolitics do not necessarily exemplify movement from primitive to advanced forms of governance, or from malevolent to benevolent, and so forth. Nor are they final.

On the other hand, such a Whiggish approach is mostly valid in the domain of science. Readers at this point may well be tired of hearing about the Copernican, sun-centered solar system, but it deserves repetition, and not only because it contributed mightily to some much needed anthropo-diminution but also because it is quite simply a better model than the Ptolemaic, Earth-centered version, just as evolution by natural selection is superior to special creation. It seems likely, by the same token, that the new paradigms of human nature described in Part II of this book are superior—more accurate, more useful, more supported by existing facts, and more likely to lead to yet more insights—than are the old paradigms they are in the process of displacing. Paul Valery wrote that a work of art is never completed but rather abandoned. This sense of unfinishedness is even more characteristic of science, not just because of the scientific "creator's" persistent personal dissatisfaction with her product but also because reality is tricky, sometimes leaving false trails. As a result, our pursuit and even our successes can always be improved on.

This is not to say (as the more extreme postmodernists claim) that reality is socially constructed and as a result, no scientific findings can be described as true. The physicist David Deutsch, making a bow to the philosopher of science Karl Popper, refers to the accumulation of scientific knowledge as "fallibilism," which acknowledges that no attempts to create knowledge can avoid being subject to error. His point is that although we may never know for sure that we are correct in an assertion (after all, tomorrow's discovery may reveal its falsity), we can be increasingly clear that we aren't completely wrong. Moreover, science has never been superseded by any better way of making sense of nature.

Newton's laws of motion are as close to truth as anyone can imagine, with respect to medium-sized objects moving at middling speeds, but even these genuine laws have been improved on when it comes to very small things (quantum physics), very fast movement (special relativity), or very large things (general relativity). Even in Newton's day, his work on gravity was vigorously resisted—by well-informed scientifically minded colleagues—because it posited pushing and pulling by a mysterious, seemingly occult "action at a distance" that could not be otherwise identified or explained. Scientific insights, in short, can be as tricky as reality. Observing the disorienting aspects of quantum physics and how they often depart from both common sense and certain rarely questioned philosophical assumptions (such as cause and effect), Steven Weinberg wryly noted, "Evidently it is a mistake to demand too strictly that new physical theories should fit some preconceived philosophical standard."[3] He might have included new biological theories as well.

Biology, in fact, offers materialist insights that, in a sense, exceed even those of physics, the presumed zenith of hard-headed objective knowledge. Thus, although physicists can tell us a lot about how matter behaves, it is so far stumped when it comes to the question of what matter is (wave functions, clouds of probability, gluons and muons and quarks and the like: but what are *they* made of?). By contrast, biology can and does tell us both what we are made of, at least at the organismic level—organs, tissues, cells, proteins, carbohydrates, fats, and so forth—and, increasingly, how and why we behave as we do: some of the deepest and hitherto unplumbed aspects of human nature.

By an interesting coincidence, in 1543, the same year that Copernicus's book on the solar system appeared, Vesalius published his magnum opus, on human anatomy. Anatomies large and small, far and near, planetary and personal: all give way to a more accurate, and typically more modest view of reality, including ourselves. Such a perspective on human behavior—mostly due to Darwin and his interpreters and extenders—can be seen as yet another continuation of this trend, dissecting the anatomy of human nature itself.

It will also become apparent in Part II of this book that biology's insights into human nature don't necessarily offer a smoother ride than does physics, because sometimes these revelations have the disconcerting habit of going counter to common sense as well as against widespread preference.

They are based, after all, on the underlying assumption (itself abundantly supported by biological facts) that we are animals, not fundamentally different from other organisms. Let's return to when Carl Sagan captured the imagination—scientific and otherwise—of millions by pointing out, in his immensely popular television series *Cosmos*, that we are "made of star stuff." He might have brought this message closer to home had he emphasized that we are also made of plant and animal stuff, which, for many, is the stuff not of starry-eyed dreams but of down-to-earth nightmares, and not only for religious fundamentalists.

Here is Robert Frost, often seen as nature-friendly, in his poem "The White-Tailed Hornet":

> As long on earth
> As our comparisons were stoutly upward
> With gods and angels, we were men at least,
> But little lower than the gods and angels.
> But once comparisons were yielded downward,
> Once we began to see our images
> Reflected in the mud and even dust,
> 'Twas disillusion upon disillusion.
> We were lost piecemeal to the animals,
> Like people thrown out to delay the wolves.

Not common sense, but "comparisons yielded downward" are, for many, the problem. Walter Benjamin—who was, if anything, nature-*un*friendly—gave voice to an additional fear: that by looking for human nature in animal nature, we weren't merely "lost piecemeal to the animals," but literally incorporated into them. "In an aversion to animals," wrote Benjamin, "the predominant fear is of being recognized by them through contact. The horror that stirs deep in man is an obscure awareness that in him something lives so akin to the animal that it might be recognized."[4] In Part II of the present book, we recognize something not only akin to animals but also, in a literal sense, indistinguishable. For those people, and there are many, who resent comparisons yielded downward, I suggest that instead of "downward," think "laterally" and thereby extended rather than shortened. Along with puncturing our species-wide delusions of grandeur, there is Darwin's famous observation that there is "grandeur in this view of life," as well as this joyous suggestion by Coleridge, from his poem "Religious Musings":

> 'Tis the sublime of man,
> Our noontide Majesty, to know ourselves
> Parts and proportions of one wondrous whole!
> This fraternizes man, this constitutes
> Our charities and bearings.

The twentieth-century physician and essayist Joseph Wood Krutch wrote movingly about what he considered a terrible loss: the fact that humanity's self-perception had degenerated from Hamlet's admiring "how like a god!" to Pavlov's matter-of-fact "how like a dog."[5] This, in turn, led B. F. Skinner to protest that the latter is actually a step forward, opening the door to verifiable scientific insights about human nature, insights that were unavailable so long as we kept perceiving ourselves as godlike, separate from the material world, and thus, mystically incomprehensible.

And so, dear reader, as you prepare to fraternize—intellectually at least—with the rest of the natural world, to know yourself "parts and proportions" of one wondrous whole, I hope you will see biology's recent insights as an array of sublime and majestic realizations, parts of a paradigm shift that both informs and enhances all of us.

NOTES

1. Mathieu Ricard, *A Plea for the Animals* (Boulder, CO: Shambala, 2016).
2. Francis Fukuyama, *The End of History and the Last Man*, reissue ed. (New York: Free Press, 2006).
3. S. Weinberg, "The Trouble with Quantum Mechanics," *New York Review of Books*, January 19, 2017.
4. Walter Benjamin, *One Way Street and Other Writings* (New York: Verso, 1997).
5. Joseph Wood Krutch, *A Krutch Omnibus: Forty Years of Social and Literary Criticism* (New York: William Morrow, 1970).

9

Uniquely Thoughtful

EVER SINCE DESCARTES, scientists have been taught and have dutifully announced that animals aren't capable of complex thought.[1] Like Gertrude Stein's famous observation about Oakland, those who deny animals any intellect or subjective mental life maintain, in effect, "there isn't any there there." But the new field of "cognitive ethology" has been revealing astounding feats of animal intelligence strongly hinting at various degrees of consciousness. Holdouts keep insisting that no other living things have a mind quite like *Homo sapiens*, and in a literal sense, they are correct. But by the same token, no other living things have a skeleton quite like ours—but that doesn't mean that they don't have their own internal bones. Looking at parrots, crows, dogs, and chimpanzees (these are the species for which data are especially convincing, although they don't exhaust the list) there is no doubt that other animals think differently from human beings, but the evidence is now irrefutable that they indeed think. And feel, and suffer.

Even before Descartes, people—at least in the West—prided themselves in being unique in the organic world, and not only in somehow possessing a divine spark. Other living things, it has been widely assumed, lack not only a soul, but also our capacity for rational thought. Recent discoveries, including work on Alex the African gray parrot and several different remarkable studies on cognition in chimpanzees, crows, and dogs, have shown that these creatures are capable of intellectual feats that compare favorably with those of normal, healthy human beings. Thus, in the realm of intellectual accomplishment, our widespread species narcissism is rapidly being dispelled.

But first, let's consider a subversive view of intelligence itself, memorably expressed by the paleontologist Jack Sepkoski: "I see intelligence as just one of a variety of adaptations among tetrapods for survival. Running fast in a herd while being as dumb as shit, I think, is a very good adaptation for survival."[2] Seen through the eyes of evolution, intelligence isn't necessarily that special. There are many ways to achieve biological success—that is, adaptations for survival—of which intellect is merely one. Consider, for example, the capacity to deal with severe drought, super high or low temperatures, extremely acidic or alkaline conditions, and so forth—in short, you could be an extremophile, which would serve you much better than having a high IQ if you were faced with living in a subterranean vent or on a Himalayan snowfield at an elevation of 30,000 feet. Or alternatively, what's wrong with covering your body with a hard shell and outfitting yourself with a pair of impressive pinchers like a king crab, or being one of a hundred thousand nonreproductive, instinct-driven workers in an anthill, so long as the queen, your mother, makes lots of babies, thereby assuring your own reproductive success by proxy?

A focus on intelligence may therefore be uncalled for—except if you happen to be a highly intelligent creature lacking extremophile adaptations, a hard shell and powerful pinchers, or the ant colony's companionate solace and are attempting to use your intelligence to identify a reason why you are so extra special. As Ecclesiastes reminds us (3:19, New American Standard Bible): "Therefore the death of man, and of beasts is one, and the condition of them both is equal: as man dieth, so they also die: all things breathe alike, and man hath nothing more than beast: all things are subject to vanity." Few things better serve our species-wide vanity than an obsession with intelligence. Or, to be more specific, with rationality. The resulting mantra is simple: rewriting Hilaire Belloc's nineteenth-century jingoistic jingle, "Whatever happens, we have got / rationality and they have not." Belloc's little rhyme celebrated not rationality but the Maxim gun, the first recoil-operated machine gun, invented by Hiram Stevens in 1883 and widely perceived to be the weapon most responsible for British imperial conquest. Rationality, it can be argued, was the weapon most responsible for humanity's imperium over the planet Earth. We have it, and "they" don't.

As it turns out, this is wrong, times two: first, many of "them" are actually capable of remarkable feats of high-quality intellectual accomplishment, and second, our own species isn't nearly as rational as many of us like to think, or—when thought fails us—imagine.

There have been many efforts by *Homo sapiens* to distinguish themselves from other animals, giving rise to a long list of alternative taxonomic monikers. Here is just a partial list: *H. absconditus* (the inscrutable), *H. adorans* (the worshipper), *H. aestheticus* (the art appreciator), *H. amans* (the loving), *H. avarus* (the greedy), *H. degeneratus* (the degenerate), *H. domesticus* (the domesticated), *H. economicus* (the self-interested), *H. ethicus* (the ethical), *H. faber* (the builder, and the tool-maker), *H. generosus* (the generous), *H. imitans* (the imitative), *H. inermis* (the helpless), *H. investigans* (the curious), *H. faber* (the builder), *H. loquens* (the talker), *H. loquax* (the chatterer), *H. ludens* (the playful), *H. mendax* (the liar), *H. necans* (the killer), *H. neophilus* (the novelty-loving), *H. patiens* (the suffering), *H. pictor* (the artist), *H. poetica* (the poet), *H. prospectus* (the future planner), *H. reciprocans* (the reciprocator), *H. ridens* (the laugher), *H. sentimentalis* (the sentimental), *H. socius* (the sociable), *H. technologicus* (the technological), and *H. viator* (the pilgrim). It was Linnaeus, reflecting the eighteenth-century Enlightenment fondness for rationality, who dubbed us *H. sapiens* (the wise), which—wisely or not—has been retained, if only as testament to our continuing self-esteem when it comes to thinking about our thinking.

Roughly 1500 years earlier, the neo-Platonic philosopher Porphyry (third century AD) had made a valiant effort to separate us from other creatures, ultimately settling—logically enough, it appears—on our rationality, the presumed secret sauce that makes us human and thus, different from other, "lower" animals:

> Man is a substance; but because an angel is also a substance, that it may appear how man differs from an angel, substance ought to be divided into corporeal and incorporeal. A man is a body, an angel, without a body: But a stone also is a body: that therefore a man may be distinguished from a stone, divide bodily or corporeal substance into animate and inanimate, that is, with or without a soul. Man is a corporeal substance animate. But plants are also animate. Let us divide therefore again corporeal substance animate into feeling and void of feeling. Man feels, a plant not: But a horse also feels, and likewise other beasts. Divide we therefore animal corporeal feeling substance into rational and irrational. Here therefore are we to stand, since it appears that every, and only man is rational.

Regrettably for Porphyry, if not for the rest of us, it isn't true that every man (or every human being) is rational. Nor are human beings uniquely so. Hence, the present chapter.

IT SHOULDN'T BE surprising that other animals think. After all, the common-denominator, bottom-line, take-home message of evolution by natural selection is, in a word, continuity. Hence, it is only to be expected that nearly all of those traits enlisted as possible defining human characteristics have turned out, on closer study of other animals, to be found in one or more species. Otters and elephants play, chimpanzees not only use tools but also make them, parrots and dogs reveal complex cognition; it would have been astounding if human beings were alone in the organic world when it comes to such actions. We are unique, of course, in our ability to program computers, just as Clark's nutcrackers are exceptional in their ability to remember the precise location of more than a thousand nuts that they have stashed. (Can you remember what you had for breakfast yesterday? Or where you parked your car when you come out of the supermarket?) For anthropocentric people-firsters, reports of animal cognition nonetheless constitute a paradigm that is vigorously resisted even today—especially by religious fundamentalists, eager to see in human thinking a distinctly nonanimal trait that, unlike the supposed soul, can be measured and thus gestured toward as evidence of humanity partaking of something that suggests the divine.

Surprisingly, perhaps, they haven't been alone. Although all serious scientists have long been Darwinians with respect to the human body, many have remained oddly creationist when it comes to the human mind. Even many scientists have accordingly long hesitated to acknowledge the reality of animal minds, although not primarily because of theological influence. Rather, it was a reaction against anthropomorphism ("human shape"), the widespread layperson's temptation to attribute human traits and motivations to nonhuman animals, which gave rise to notions that ants are industrious, lions are noble, camels are supercilious, owls are wise, and so forth. Anthropomorphism became an unconscious habit, and a lazy one at that, permitting observers to substitute a simplistic explanation where serious, objective research is called for. Reacting to the assertion that pigs postulate, fish philosophize, and, for all one can say, rhododendrons ruminate, scientists went "whole hog" in the other direction, such that it now seems likely that in the process, human minds were closed to the mental continuity that links human and nonhuman animals, in the course of which harm of several sorts was done.

For one thing, a nonexistent gulf was established between *Homo sapiens* and other living things, which is not only objectively false, but hurtful in that it facilitates the abuse of animals. After all, if they are mere automatons, devoid not only of a soul but of a subjective, internal mental life, then there is little to prevent them from being cruelly abused, and we should no more feel concerned about abusing a dog, cat, parrot, or elephant than a bicycle or a fire hydrant. The good news is that this error is rapidly being corrected, leading to the interesting situation whereby science comes eventually to catch up with something long acknowledged by much of the untutored public. Or as Kenneth Grahame wrote in *The Wind in the Willows*,

> The clever men at Oxford
> Know all that there is to be knowed.
> But they none of them know one half as much
> As intelligent Mr. Toad.

FIRST, A QUICK look at how the erroneous view that other animals lack an internal mental life got started. It wasn't merely a consequence of self-serving theology and a reaction to anthropomorphism, but has some legitimate basis in science. Early research by ethologists focused on the role of "releasers," simple stimuli that are typically present in members of a given species, to which individuals respond automatically and instinctively—nearly always without any indication of insight. The classic example was the orange spot near the tip of an adult herring gull's bill. Niko Tinbergen (who shared a Nobel Prize for his groundbreaking research) found that herring gull chicks instinctively pecked at this spot, presented by a parent gull, who then responded—also instinctively—by regurgitating a semidigested fish meal into the mouth of the hungry chick. Especially noteworthy is that chicks were equally likely to peck at something that at least to the human observer doesn't look at all like the head of a herring gull, namely a tongue depressor or Popsicle stick on which an orange dot has been painted.

Tinbergen also noticed that male stickleback fish, kept in his lab at Oxford, used to rush to the side of their aquarium and display aggressively when mail trucks—painted red in England at the time—drove by. Not coincidentally, male sticklebacks develop bright red chests when in breeding condition, and their cognitive skills are so limited (or more accurately, in this case, bypassed by a simple algorithm) that they take anything moving and bright red as a signal that automatically releases aggressive territorial defense. In my own research, I have taken advantage of a similar releaser among male mountain bluebirds, whose bright blue plumage can be mimicked by a bright blue racquetball, to the extent that the ball, impaled on a nearby tree, evokes a full repertoire of mountain bluebird display behavior. Such examples are common currency among students of animal behavior.

Further evidence for the cognitive limitations of many animals comes from the existence of "supernormal releasers," created when researchers take a releaser and enhance or exaggerate it to supernormal proportions, whereupon it evokes a supernormal response, demonstrating once more—often to comical excess—an absence of deep thought (or even shallow thought) among a wide range of animals. A notable example is provided by Pacific oystercatchers, rather debonair looking shorebirds with black and white plumage, and bright orange legs and beak. These crow-sized birds lay eggs that are lightly speckled and appropriate for their mass—a bit smaller than hen's eggs—and they incubate them. Take a watermelon, however, and paint it with similar speckling, and the seemingly addle-pated oystercatcher will abandon her own eggs and perch, apparently quite satisfied, albeit looking entirely absurd to a human observer, atop this supernormal releaser (which might weigh twenty times her body mass and could never have been laid by the animal in question), all the while ignoring her own eggs.

Put these observations together; add the fact that the animals typically studied by the classical ethologists such as Tinbergen and Konrad Lorenz have been insects, fish, and birds; sprinkle in scholarly resistance to the folk diagnosis of anthropomorphism; and then stir carefully with the intellectual spice deriving from ethologists' reaction to the traditional emphasis by comparative psychologists on animal learning, and the conclusion that animals lack complex cognition becomes almost unavoidable. The final coup de grace came from Occam's Razor, the scientific principle that natural explanations should not be elaborated unnecessarily—that is, when in doubt, the simplest, least elaborate explanation, requiring the smallest number of additional assumptions, should be taken as correct.

As it happens, Occam's Razor is generally a good rule of thumb. But it is neither sacrosanct, nor so sharp that it cuts through all aspects of reality. There is nothing inherently valid about simplicity; sometimes, the nature of the world is best explained by bafflingly complex laws and patterns. (Just take a look at the equations in a physics textbook, especially a graduate-level treatment.) By the same token, releasers and supernormal releasers often—if not "generally"—exist among animals, especially those with relatively simple brains whose behavior is most efficiently tuned to provide equally simple reactions to simple stimuli. A healthy reaction to this minimalist perspective on animal behavior has however been gaining adherents in recent years. Known as "cognitive ethology," or "evolutionary cognition," it has effectively broken the long-standing taboo against giving animals their due, acknowledging that for many species, and in a variety of circumstances, animals experience a rich mental world.

The biologist Marc Bekoff and primatologist Frans de Waal have recently been especially effective in puncturing the myth of animal mental vacancy and, in the process, giving the lie to the myth that only human beings think. Most influential in this respect was the biologist Donald Griffin; and herein lies an interesting story, not only of science but also of the sociology of researchers. For decades, the study of animal cognition (and a related, even more controversial assertion, animal consciousness) was the third rail of ethological research: touch it and albeit you wouldn't be electrocuted, you certainly wouldn't get a research grant, or tenure. Professor Griffin was no starry-eyed animal lover, addicted to anecdotes about his pet cat; rather, he was a highly regarded scientist, who, through a series of careful empirical studies had discovered that bats used echolocation (essentially, a form of ultra-high-frequency sonar) to avoid obstacles and detect their insect prey. Griffin then shocked the animal behavior establishment when in a series of carefully argued books—notably *Animal Minds*, *Animal Thinking*, and *The Question of Animal Awareness*—he urged his colleagues to take seriously the questions of, well, animal minds, animal thinking, and animal awareness.

Once a giant like Dr. Griffin said it was kosher, a growing number of biologists and psychologists began treading where no self-respecting scientist had previously allowed herself to go. The results have been overwhelmingly persuasive, such that these days, virtually no scientist publicly doubts the mental life of nonhuman animals.

Let's start, not surprisingly, with chimpanzees. In a report published in the journal *Science*, three researchers at the Max Planck Institute for Evolutionary Anthropology, in Leipzig, reported on chimps exposed to two slightly different experimental setups. In one, the apes were able to obtain food by their solo efforts; in the other, they needed the assistance of a second individual. The subject animal could recruit another by literally taking a key, unlocking a door, and releasing the potential assistant. When no collaboration was needed, subjects did not open the door; hence, they obtained the food without being obliged to share. When the coordinated effort of two chimps was required in order to get the food, subjects obtained help. Of special interest—and difficult to interpret without allowing them considerable intellectual acumen—is this finding: when subjects were given the opportunity to choose between two different potential collaborators, kept in adjacent rooms, and when the subjects had been given prior experience in working with each of the two, they preferentially released the one who had previously shown herself to be a more effective collaborator.[3]

There are at least two complex, interconnected parts to the mental agility herein demonstrated: First, recognizing when collaboration is and is not necessary and responding appropriately in each case. Second, ranking the collaborative value of different individuals as a result of previous interactions, remembering their relative value, and also responding differentially and appropriately to those individuals when the circumstance called for doing so.

Here is another remarkable example of chimpanzee intellect: A subordinate chimp witnessed two pieces of food being hidden; in one case, a dominant chimp had also been watching, whereas in the other, the subordinate was the only observer. Later, when given the opportunity, the subordinate avoided the food that the dominant knew about, and took only the "safe" morsel, which didn't evoke the dominant's ire. It is an interesting exercise to run through, in your own human mind, the intellectual process that must have gone through that of the subordinate chimp.[4]

Here is yet another interesting exercise. We share about 99 percent of our genome with chimps and bonobos. Isn't it possible that somewhere in the universe there exist creatures who—perhaps by virtue of simply differing in their genetic code (or equivalent) by as little as another 1 percent—would look at us as being as much beneath them, intellectually, as so many humans perceive other animals?

Nor is humanlike intelligence limited to our great ape cousins. Consider the case of Rico, a nine-year-old border collie from Dortmund, Germany, whose mental gymnastics were also reported in *Science*, once again from Leipzig's Max Planck Institute for Evolutionary Anthropology. By the time of the study, Rico had acquired a vocabulary of more than two hundred words, most of them nouns, which he demonstrated by successfully retrieving thirty-seven of forty toys from a collection in a different room, when its name was called by his owner/interlocutor.

In perhaps the most notable test, Rico showed an impressive command of logic. Seven toys, each of whose names Rico knew, were placed in a separate room along with an eighth, which was new to the dog. He was then commanded to return with this eighth item, the name of which he had not previously heard. Seven out of ten times, Rico came back with this eighth, previously unknown object. Presumably, his internal dialog went something like this: "I'm supposed to fetch the widget, whatever that is. Here are seven things, each of which I know; and none is a widget. So, this eighth one must be it." Four weeks later, Rico was able to remember the item three out of six times, comparable to the performance of three-year-old children. Of course, border collies have been bred for close collaboration with human beings, and it remains to be seen whether, for example, Rico can learn *not* to fetch an object on command.[5]

Nonetheless, "for psychologists," wrote Yale's Paul Bloom in an accompanying article in the same issue of *Science*, "dogs may be the new chimpanzees."

One might suppose that these intellectual feats are the province of mammals only, with their relatively large brains. This conceit is summarily shattered by the mental exploits of Alex, an African gray parrot who was studied for three decades by Irene Pepperberg, currently at Brandeis University. Alex, whose brain is smaller than a peanut, proved to be astoundingly adroit, suggesting that birds may be the new dogs.

Recounted in Pepperberg's book, *The Alex Studies*, as well as in numerous filmed appearances, Alex had more than fifty words for different objects. He could also name seven colors (Rico, for all his evident brilliance as a listener/responder, isn't much of a conversationalist) and five different shapes, could count to six, and (perhaps most remarkably) was able to combine these ideas . . . and they are ideas, not just words for objects, in meaningful ways. Dr. Pepperberg maintains that Alex had the language abilities of a two-year-old child, and the problem-solving skills of a six-year-old.

All of which would seem to confirm the suggestion by John Morley, in his *Life of Gladstone*, that "Simplicity of character is no hindrance to subtlety of intellect"[6]— except that those people who have been kept by African gray parrots may well reply that the characters generated by these supposed birdbrains are far from simple!

Given a tray with wooden blocks and wool balls of different colors, and asked "On the tray, how many orange wool?" Alex responded, correctly, "Four."[7] Think about it (Alex evidently did): he not only had to discriminate wool from wood, and know which word refers to which, but also distinguish the differently colored wool balls, as well as count them, all in response to one complex request.

When shown an array of things, Alex responded with greater than 80 percent accuracy to questions such as "What object is green and three-corner?" or "What color is wood and four-corner?" or "What shape is paper and purple?" He also understood the concept "different," being able to pick out, on command, the different one from an array of four things of which three are, for example, of the same color, or three are large and one is small. Recall the Sesame Street song, "One of These Things Is Not Like the Other," designed to hone the reasoning skills of human preschoolers.

Now that the dam against studying animal intelligence has finally been breeched, a veritable flood of research findings has been emerging. In his book *Are We Smart Enough to Know How Smart Animals Are?*,[8] the primatologist Frans de Waal presents an encyclopedic catalog of animal "smartness," leading to the irrevocable conclusion that human beings do not have a monopoly on tool use, toolmaking, empathy, self-recognition, a sense of fairness, cooperative behavior, mental time-travel, rudimentary language, imitative culture, and numerous other "higher" abilities and traits.

Quoting the zoologist Desmond Morris, de Waal describes a daily ritual that used to occur at the London Zoo, when teacups, a teapot, and dainty dishes and chairs were set out for the chimpanzees to wreak havoc and amuse the onlookers. Soon enough, a problem arose: they (the chimps) quickly learned how to take tea like Victorian gentlefolk, and so, de Waal writes, "When the public tea parties began to threaten the human ego, something had to be done. The apes were retrained to spill the tea, throw food around, drink from the teapot's spout,"[9] and so on. The animals, in short, had to be *taught* to be as foolish and ill-behaved—in a word, stupid—as the public had assumed they were.

Answering his question, de Waal writes, "yes, we are smart enough to appreciate other species, but it has required the steady hammering of our thick skulls with hundreds of facts that were initially poo-pooed by science."[10] Human cognition, it is increasingly clear, is really just a variety of animal cognition. The mental capacity of each animal species is tuned by natural selection to the kinds of situations encountered by each species, whether opening nuts, remembering the location of cached food, keeping track of social relationships, distinguishing poisonous from nutritious food, and so on. It isn't a matter of smart versus stupid, but rather, given the anatomy, physiology, DNA, and experiences and opportunities characteristic of each species, what are the most effective intellectual tools? Different mental strokes for different biological and psychological folks.

Caution is nonetheless called for when assessing claims of animal cognitive skills that appear remarkable to the human observer. It is one thing to be generous in interpreting the behavior of other animals, quite another to be taken in. Students of animal behavior are still smarting after having been overly credulous about the intellect of Clever Hans, a reputedly brilliant horse. Beginning in the early 1890s, William von Osten astounded European audiences with his horse's ability to answer difficult questions, involving, for example, arithmetic. Asked for the sum of three and two, Hans demonstrated his cleverness by tapping his hoof five times. And since it wasn't necessary for von Osten to be present, it seemed most unlikely that any trickery was involved.

But eventually, it was realized that whenever no one in the room knew the answer to a question posed to him, Clever Hans didn't know, either. It turns out that he had learned to respond—by ceasing his hoof tapping—to very subtle nonverbal cues provided by the human beings around him. People would unintentionally tense their muscles until Hans reached the right answer; then they would relax and Hans, sensing this, stopped moving his hoof. Hans was indeed clever, but not as advertised.

These days, researchers in animal behavior are clever, too, and in the studies recounted here—as well as in numerous others accumulating in the scientific literature—they took pains to avoid any hint of unconscious cuing.

At the same time, it is now widely understand that sauce for the human goose also works for the animal gander: intelligence may be mysterious, in the sense of difficult to unravel, but it is no more mystical than any other property of living things. The pioneering geologist Sir Charles Lyell made an intriguing suggestion when he wrote, in *The Geological Evidences of the Antiquity of Man* (1863), that

> so far from having a materialistic tendency, the supposed introduction into the earth at successive geological periods of life—sensation, instinct, the intelligence of the higher mammalia bordering on reason, and lastly, the improvable reason of Man himself—presents us with a picture of the ever-increasing dominion of mind over matter.

BUT NOTWITHSTANDING HIS formative influence on Darwin's evolutionary insights, in this instance Lyell was wrong. The question isn't a continuous trajectory of "mind over matter" but rather one of *mind deriving from matter*, whereby the matter of animal brains seems no less capable of generating mind than is the matter of which human brains are composed. The differences are of degree, not kind.

The question of animal awareness and consciousness isn't a question of demoting our species, but of seeing ourselves as part of a wider and deeper natural continuum. It also carries some additional freight, which isn't explored here but deserves mention: the question of how we treat our fellow minded creatures. As the philosopher Peter Singer pointed out in his aptly named book *Animal Liberation*, "ending oppression and exploitation wherever they occur, and in seeing that the basic moral principle of equal consideration of interests is not arbitrarily restricted to members of our own species."[11]

In evolutionary context, intelligence is a biological strategy whereby organisms, human or nonhuman animal, possess sufficient behavioral complexity and flexibility to respond adaptively to complex situations. Under some conditions, individuals with such qualities were simply more fit than those lacking them, thereby selecting for differing levels of intelligence in certain species, just as natural selection has favored differing patterns of cell metabolism, kidney filtration, or blood circulation.

There is even a small but growing cadre of botanists who argue for the legitimacy of the new field of "plant neurobiology." Plants certainly behave, responding appropriately and integratively to diverse stimuli. In the process, they even make use of electrical and hormonal communication. Whether they are "intelligent" or even "conscious" is another matter, but probably not one that should immediately be foreclosed. After all, cucumbers may be the next parrots.

"Almost every week there is a new finding regarding sophisticated animal cognition, often with compelling videos to back it up," writes Frans de Waal. He describes the rapidly accumulating data showing that rats may regret their own decisions, that crows manufacture tools, that octopuses recognize human faces, and that special neurons allow monkeys to learn from each other's mistakes. We speak openly about culture in animals and about their empathy and friendships. Nothing is off limits anymore, not even the rationality that was once considered humanity's trademark.[12]

Time to carry this observation one step farther. Not only is animal rationality no longer off limits but also it is well established. At the same time, it is increasingly clear that rationality isn't necessarily our trademark, because to a disconcerting extent, we aren't that rational after all.

"What a piece of work is a man!" exulted Shakespeare's otherwise melancholy Hamlet. "How noble in reason! How infinite in faculty! In form and moving how express and admirable! In action how like an angel! In apprehension how like a god!" Nearly two thousand years earlier, Aristotle maintained that happiness comes from the use of reason, since that is the unique glory and power of humanity. Indeed, for "the Greeks" generally, reason distinguishes us from all other living things, and the life of reason is thus the greatest good to which human beings can aspire. So why doesn't it attract more adherents these days?

For one thing, it may simply be that reason—by definition—is dry and cerebral, only rarely making inroads below the waist. Omar Khayyam made this trade-off uniquely explicit: "For a new Marriage I did make Carouse: / Divorced old barren Reason from my Bed / And took the Daughter of the Vine to Spouse."

To be sure, excessive reason is easy to caricature. Thus, at one point in Jonathan Swift's *Gulliver's Travels*, our hero journeys to Laputa, whose (male) inhabitants are utterly devoted to their intellects: one eye focuses inward and the other on the stars. Neither looks straight ahead. The Laputans are so cerebral that they cannot hold a normal conversation; their minds wander off into sheer contemplation. They require servants who swat them about the mouth and ears with special instruments, reminding them to speak or listen as needed. Laputans concern themselves only with pure mathematics and equally pure music. Appropriately, they inhabit an island that floats, in ethereal indifference, above the ground. Laputan women, however, are unhappy and regularly cuckold their husbands, who do not notice. The prime minister's wife, for example, repeatedly runs away, preferring to live down on Earth with a drunk who beats her.

Thus presented, to reject reason actually seems downright reasonable! Consider how rare it is for someone caught in the grip of strong emotion to be overcome by a fit of rationality, but how frequently events go the other way. After all, Blaise Pascal, who abandoned his brilliant study of mathematics to pursue religious contemplation, famously noted, "the heart has its reasons that reason does not understand." Or as the seventeenth-century English churchman and poet Henry Aldrich pointed out in his *Reasons for Drinking*, often we make up our minds first, and find "reasons" only later:

> If all be true that I do think
> There are five reasons we should drink:
> Good wine—a friend—or being dry—
> or lest we should be by and by—
> Or any other reason why!

We may speak admiringly of Greek rationality, of the Age of Reason, and of the Enlightenment, yet it is far easier to find great writing—and even, paradoxically, serious thinking—that extols unreason, irrationality, and the beauty of "following one's heart" rather than one's head. Some of the most "rational" people have done just that.

Legend has it, for example, that when Pythagoras came up with his famous theorem, justly renowned as the cornerstone of geometry (that most logical of mental pursuits) he immediately sacrificed a bull to Apollo. Or think of Isaac Newton: pioneering physicist, both theoretical and empirical, he of the laws of motion and of gravity, inventor of calculus, and widely acknowledged as among the greatest of all scientists, perhaps *the* greatest. This same brilliant intellect wrote literally thousands of pages, far more than all his physics and mathematics combined, seeking to explicate the prophecies in the Book of Daniel.

Montaigne devoted many of his essays to a skeptical denunciation of the human ability to know anything with certainty. But probably the most influential of reason's opponents was Jean-Jacques Rousseau, nineteenth-century philosopher and social commentator who claimed in his *Second Discourse* that "the man who thinks is a depraved animal," thereby speaking for what came to be the Romantic movement. But even earlier, many thinkers, including those who employed reason with exquisite precision, had been inclined to put it "in its place." The hard-headed empiricist philosopher David Hume, for example, proclaimed in his *Treatise of Human Nature* that "reason is, and ought to be, the slave of the passions, and can never pretend to any other office than to serve and obey them." Furthermore, when reason turns against the deeper needs of people, Hume argued, people will turn against reason.

Perhaps the most articulate, not to mention downright angry, denunciation of human reason, however, is found in the work of Fyodor Dostoyevsky, especially his novella *Notes from Underground*, which depicts a nameless antihero: unattractive, unappealing, and irrational (although intelligent!). In angry contradiction to the utilitarians who argued that society should aim for the "greatest good for the greatest number" and that people can be expected to act in their own best interest, the Underground Man—literature's first "antihero"—jeered that humanity can never be encompassed within a "Crystal Palace" of rationality. He may have a point: certainly, unreason can be every bit as "human" as the Greeks believed rationality to be. You don't have to be a Freudian, for example, to recognize the importance of the unconscious, which, like an iceberg, not only floats largely below the surface—and is thus inaccessible to rational control—but also constitutes much of our total mental mass.

It is one thing, however, to acknowledge the importance of unreason and irrationality, and quite another to applaud it, as the Underground Man does: "I am a sick man . . . . I am a spiteful man. I am a most unpleasant man." The key concept for Dostoyevsky's irrational actor is spite, a malicious desire to hurt others—including one's self—without any compensating gain for the perpetrator. Consider the classic formulation of spite: "cutting off your nose to spite your face," disfiguring yourself for "no reason."

Significantly, spiteful behavior does not occur among animals. Even when an animal injures itself or appears to behave irrationally—gnawing off its own paw, killing and eating its offspring—there is typically a biological payoff: freeing oneself from a trap, turning one or more offspring (who under certain circumstances may be unlikely to survive) into calories for the parent. Spite, by contrast, is uniquely human.

The Underground Man goes on to rail against a world in which—to his great annoyance—two times two equals four. He claims, instead, that there is pleasure to be found in a toothache, and refers, with something approaching admiration, to Cleopatra's alleged fondness for sticking golden pins in her slave-girls' breasts in order to "take pleasure in their screams and writhing." As the Underground Man sees it, the essence of humanness is living "according to our own stupid will . . . because it preserves for us what's most important and precious, that is, our personality and our individuality." He believes that people act irrationally because they stubbornly want to, snarling, "If you say that one can also calculate all this according to a table, this chaos and darkness, these curses, so that the mere possibility of calculating it all in advance would stop everything and that reason alone would prevail—in that case man would be insane deliberately in order not to have reason, but to have his own way!"

Such sentiments are in no way limited to this most famous apostle of the dark Russian soul, or to European Romantics. Here is a poem ("In the Desert") from that quintessentially American writer, Stephen Crane, who gave us *The Red Badge of Courage*:

> In the desert
> I saw a creature, naked, bestial,
> Who, squatting upon the ground,
> Held his heart in his hands,
> And ate of it.
> I said, "Is it good, friend?"
> "It is bitter—bitter," he answered;

> "But I like it
> Because it is bitter,
> And because it is my heart."

No matter how fashionable it may be to "dis" reason, let's not be carried away. (By what? Presumably, by unreason, since as already suggested, people aren't generally swept away in an uncontrollable fit of rationality.) Strong emotion can be wonderful, especially when it involves love. But it can also be horrible, as when it calls forth hatred, fear, or violence. In any event, one doesn't have to idolize Greek-style rationality to recognize that excesses of unreason typically have little to recommend themselves, and much misery to answer for.

We may admire—albeit surreptitiously—the Underground Man's insistence on being unpredictable, even unpleasant, spiteful, or willfully irrational. But most of us wouldn't choose him to be our financial, vocational, or romantic adviser, or, indeed, any sort of purveyor of wisdom. Maybe unalloyed reason doesn't make the heart sing, but as a guide to action, it is probably a lot better than its darker, danker, likely more destructive, albeit sexier alternative.

In Newton's case, as in Pythagoras's, the most exquisite rationality did not preclude unreason—or, as some would prefer to call it, faith. But at least, no great harm seems to have been done by the cohabitation. Sadly, this isn't always the case. "Only part of us is sane," wrote Rebecca West.

> Only part of us loves pleasure and the longer day of happiness, wants to live to our nineties and die in peace, in a house that we build, that shall shelter those who come after us. The other half of us is nearly mad. It prefers the disagreeable to the agreeable, loves pain and its darker night of despair, and wants to die in a catastrophe that will set life back to its beginnings and leave nothing of our house save its blackened foundations. Our bright natures fight in us with this yeasty darkness, and neither part is commonly quite victorious, for we are divided against ourselves.[13]

It may be significant that Ms. West wrote the above while reminiscing on her time in the Balkans, among inhabitants of what we now identify as the former Yugoslavia, people with a long, terrible history of doing things to each other that many outsiders readily label "insane," or at least, "unreasonable." Her point is deeper however, not merely a meditation on Balkan irrationality, but on everyone's.

Take, for a more pedestrian example, the following, derived from the now-classic research of Daniel Kahneman and Amos Tversky: "Imagine that you have decided to see a play and paid the admission price of $10 per ticket. As you enter the theater, you discover that you have lost the ticket. The seat was not marked, and the ticket cannot be recovered. Would you pay $10 for another ticket?" Forty-six percent of experimental subjects answered yes; 54 percent answered no.

Then, a different question was asked: "Imagine that you have decided to see a play where admission is $10 per ticket. As you enter the theater, you discover that you have lost a $10 bill. Would you still pay $10 for a ticket for the play?' The results: This time, a whopping 88 percent answered yes and only 12 percent answered no. In other words, most people say that if they had lost their ticket, they would be unwilling to buy another, but if they had simply lost the value of the ticket ($10), an overwhelming majority have no qualms about making the purchase! Why such a huge difference? According to the psychologists Kahneman and Tversky (the former an economics Nobel Prize winner), it is explicable—not by reason but by the way people organize their mental accounts.

Here is another one: "Would you accept a gamble that offers a 10 percent chance to win $95 and a 90 percent chance to lose $5?" The great majority of people in the study rejected this proposition as a loser. Yet, a bit later, the same individuals were asked this question: "Would you pay $5 to participate in a lottery that offers a 10 percent chance to win $100 and a 90 percent chance to win nothing?" A large proportion of those who refused the first option accepted the second. But the options offer identical outcomes. As Kahneman and Tversky see it: "Thinking of the $5 as a payment makes the venture more acceptable than thinking of the same amount as a loss." It's all a matter of how the situation is framed, in this case, the extent to which people are "risk averse."[14]

Which brings us to yet another perspective on why *Homo sapiens* isn't always strictly sapient. Let's start by agreeing with Herbert Simon (another psychologist who won a Nobel Prize in economics, by the way) that the mind is simply incapable of solving many of the problems posed by the real world, just because the world is big and the mind is small. But add this: the human mind did not develop as a calculator designed to solve logical problems. Rather, it evolved for a very limited purpose, one not fundamentally different from that of the heart, lungs, or kidneys; that is, the job of the brain is simply to enhance the reproductive success of the body within which it resides. (And in the process, to promote the success of the genes that produced the body: brain and all.)

This is the biological purpose of every mind, human as well as animal, and moreover, it is its only purpose. The immediate purpose of the heart, on the other hand, is to pump blood, of the lungs to exchange oxygen and carbon dioxide, and of the kidneys to eliminate toxic chemicals. The brain's purpose is to direct our internal organs and our external behavior in a way that maximizes our evolutionary success. That's it. Given this, it is remarkable that the human mind is good at solving any theoretical problems whatsoever, beyond "Whom should I mate with?," "What is that guy up to?," "How can I help my kid?," "Where are the antelopes hanging out at this time of year?" There is nothing in the biological specifications for brain-building that calls for a device capable of high-powered reasoning, of solving quadratic equations, or even providing an accurate picture of the "outside" world, beyond what is needed to enable its possessors to thrive and reproduce. Put these requirements together, on the other hand, and it appears that the result turns out to be a pretty good (that is, rational) calculating device.

In short, the evolutionary design features of the human brain may hold the key to our penchant for logic as well as illogic. Following is a particularly revealing example, known as the Wason test.

Imagine that you are confronted with four cards. Each has a letter of the alphabet on one side and a number on the other. You are also told this rule: If there is a vowel on one side, there must be an even number on the other. Your job is to determine which (if any) of the cards must be turned over in order to determine whether the rule is being followed. However, you must only turn over those cards that require turning over. Let's say that the four cards are as follows:

G   4   A   7

Which ones should you turn over?

Most people realize that they don't have to inspect the other side of card "G." However, a large proportion respond that the "4" should be inspected. They are wrong: The rule says that if one side is a vowel, the other must be an even number, but nothing about whether an even number must be accompanied by a vowel. (The side opposite an "4" could be a vowel or a consonant; either way, the rule is not violated.) Most people also agree that the "A" must be turned over, since if the other side is not an even number, the rule would be violated. But many people do not realize that the "7" must also be inspected: if its flip-side is a vowel, then the rule is violated. So, the correct answer to the above Wason test is that "G" and "4" should not be turned over, but "A" and "7" should be. Fewer than 20 percent of respondents get it right.

Next, consider this puzzle: You are a bartender at a nightclub where the legal drinking age is twenty-one. Your job is to make sure that this rule is followed: People under twenty-one must not be drinking alcohol. Toward that end, you can ask individuals their age, or check what they are drinking, but you are required not to be any more intrusive than is absolutely necessary. You are confronted with four different situations, as shown below. In which case (if any) should you ask a patron his or her age, or find out what beverage is being consumed?

#1	#2	#3	#4
Drinking	*Over*	*Drinking*	*Under*
Water	*21*	*Beer*	*21*

Nearly everyone finds this problem easy. You needn't check the age of person #1, the water-drinker. Similarly, there is no reason to examine the beverage of person #2, who is over 21. But obviously, you had better check the age of person #3, who is drinking beer, just as you need to check the beverage of person #4, who is underage. The point is that this problem set—which is nearly always answered correctly—is logically identical to the earlier set, the one that causes considerable head-scratching, not to mention incorrect answers.

Why is the second problem set so easy, and the first so difficult? This question has been intensively researched by the evolutionary psychologist Leda Cosmides.[15] Her answer is that the key isn't logic itself—after all, the two problems are logically equivalent—but how they are positioned in a world of social and biological reality. Thus, whereas the first is a matter of pure reason, disconnected from reality, the second plays into matters of truth-telling and the detection of social cheaters. The human mind, Cosmides points out, is not adapted to solve rarified problems of logic, but is quite refined and powerful when it comes to dealing with matters of cheating and deception. In short, our rationality is bounded by what our brains were constructed—that is, evolved—to do.

One of Goya's most famous paintings is titled *The Sleep of Reason Produces Monsters*. Monsters, however, arise from many sources, and not just when reason is slumbering and our irrational, unconscious selves have free play. Sometimes it is reason itself that generates monstrous outcomes. After all, the gas chambers of Auschwitz and the nuclear devastation of Hiroshima and Nagasaki were technical triumphs, involving no small amount of "rationality."

It is hard not to conclude that just as complex cognition is found among many non-human animals, *Homo sapiens* can be remarkably unreasonable. We have long been in the habit of understating the originality and cognitive complexity of our animal relatives, while exaggerating our own. Time for a new paradigm.

Old paradigm: Nonhuman animals are unreasoning automatons; people, by contrast, are notable for—even defined by—their use of reason. Moreover, our species is unique in possessing an internal mental life.

New paradigm: Human beings have not cornered the market on complex cognition and an array of complex mental capacities; moreover, we are often downright irrational, and not merely when in the temporary "throes of passion" but also as part of our complicated human nature.

NOTES

1. Some material in this chapter has been revised and repurposed from my articles titled "Animal Intelligence" (http://www.chronicle.com/article/Animal-Intelligence/9433) and

"Unreason's Seductive Charms" (http://www.chronicle.com/article/unreasons-seductive-charms/32123), appearing originally in the *Chronicle of Higher Education*.

2. Quoted by Michael Ruse in *From Monad to Man: The Concept of Progress in Evolutionary Biology* (Cambridge, MA: Harvard University Press, 1997).

3. A. P. Melis, B. Hare, and M. Tomasello, "Chimpanzees Recruit the Best Collaborators," *Science* 311, no. 5765 (2006): 1297–1300.

4. B. Hare et al., "Chimpanzees Know What Conspecifics Do and Do Not See," *Animal Behaviour* 59, no. 4 (2000): 771–785.

5. J. Kaminski, J. Call, and J. Fischer, "Word Learning in a Domestic Dog: Evidence For Fast Mapping," *Science* 304, no. 5677 (2004): 1682–1683.

6. John Morley, *The Life of Gladstone*, vol. 2 (London: Macmillan, 1903).

7. Irene Pepperberg, *The Alex Studies: Cognitive and Communicative Abilities of Grey Parrots* (Cambridge, MA: Harvard University Press, 2002).

8. Frans de Waal, *Are We Smart Enough to Know How Smart Animals Are?* (New York: Norton, 2016).

9. De Waal, *Are We Smart Enough?*

10. De Waal, *Are We Smart Enough?*

11. Peter Singer, *Animal Liberation* (New York: New York Review of Books, 1990).

12. De Waal, *Are We Smart Enough?*

13. Rebecca West. *Black Lamb and Grey Falcon* (New York: Viking, 1941).

14. D. Kahneman and Amos Tversky, "Prospect Theory: An Analysis of Decision Under Risk," *Econometrica: Journal of the Econometric Society* 47, no. 2 (1979): 263–291.

15. Leda Cosmides, "The Logic of Social Exchange: Has Natural Selection Shaped How Humans Reason? Studies with the Wason Selection Task," *Cognition* 31, no. 3 (1989): 187–276.

10

Conflict between Parents and Offspring

"THE COURSE OF true love never did run smooth," as Lysander notes in Shakespeare's *A Midsummer Night's Dream*. Although describing romantic love, this observation also applies to parents and their offspring, and this, in turn, seems counter to basic evolutionary wisdom. After all, the biological interests of parents and children would appear to coincide perfectly, given that the latter are the former's most direct route to evolutionary success. The two generations ought therefore to be on the same page, because successful children literally mean successful parental genes. (Success in projecting those genes into the future—the evolutionary meaning of fitness—is, in fact, the only biological reason to reproduce in the first place.)[1]

Years of Walt Disney's *True Life Adventures,* combined with classic cartoon images from *Dumbo,* to *Bambi, Lady and the Tramp*, and *One Hundred and One Dalmatians,* have reflected, as well as generated, the expectation that animal parent and child—especially mother and child—are the epitome of shared goals and perfect amiability. The image among human beings is if anything even more clearly established: Madonna and Child convey a sense of peace and contentment that transcends the merely theological.

When rough spots emerge in the parent-child nexus, the traditional view among psychiatrists, psychologists, and sociologists—such as John Bowlby, D. W. Winnicott, Jerome Kagan, and Talcott Parsons—has long been that the culprit is simply misunderstanding, with its attendant failures of communication. Everyone means well. It's just that in the course of conveying heartfelt parental assistance, advice, protection, nurturance, and information to the child, sometimes there are problems, largely because the child—being young—is necessarily inexperienced, perhaps occasionally a bit headstrong, and generally uninformed as to its true interests, which necessarily are congruent with the parents. The more mature child gradually recognizes that it is best to go along with parental inclinations and injunctions, at which point conflict ceases and healthy "socialization" has been achieved. Parent–offspring conflict is thus largely due to the fact that children are primitive, even barbaric little creatures, who need time to grow into responsible adults.

Then there is the psychoanalytic tradition, which focuses on sexual rivalry, especially between fathers and sons. As a result of presumed Oedipal conflict, boys are terrified that their fathers will castrate them; girls, for their part, resent not having a penis, and so all hell breaks loose until, in time, things quiet down and the child achieves rational self-interest, reconciling its innate sexual conflicts to its social role. Once again, it's a matter of children growing up—emotionally and intellectually no less than physically.

The view from evolutionary biology is quite different, rather darker, and much more persuasive, exemplifying yet another case in which a modern understanding of human nature gives rise to a substantial paradigm shift.

We owe it almost entirely to Robert L. Trivers, perhaps our most creative living evolutionary theorist. It is said that when Thomas Huxley first read Darwin's *Origin of Species*, he exclaimed, "How stupid of me not to have thought of that!" Trivers has been in the habit of evoking similar responses from modern biologists, with ideas that are equally "obvious" . . . after he points them out! In a landmark analysis published in 1974,[2] Trivers pointed out that the conventional wisdom regarding parent–offspring relations was all wet.

These days, evolutionary thinking has permeated not just biological and (increasingly) social science, but popular culture as well. Surprisingly, however, Trivers's theory of why evolution creates intergenerational conflict, which has major implications for family dynamics, has largely been ignored by psychologists, psychiatrists, and sociologists.

Trivers emphasized that although parents and offspring do have a substantial shared genetic interest, the overlap is not complete. Standard diploid organisms such as human beings possess two sets of chromosomes, one set from the mother (conveyed via her egg) and one from the father (conveyed via his sperm). During oogenesis and spermatogenesis respectively, the number of chromosomes and thus the genotype of each parent is reduced by one-half from its doubled, diploid state to haploid gametes— those eggs and sperm. When these unite at fertilization to form a new individual, the diploid genotype characteristic of the parents is reestablished. If such reduction divisions, called meiosis, did not precede fertilization, genomes would be doubled every time sexual reproduction takes place, which clearly would not be a viable situation.

Another way of looking at this: sexual reproduction is structured so that there is a 50 percent probability that any gene present in a parent is also present in the child. Insofar as the child succeeds, the parent does, too. Or, at least, one-half of the parent. Things would be different if we reproduced asexually, in which case each offspring would be identical to each parent, and there would be a 100 percent likelihood of gene transmission from parent to child. But just as there are often two sides to every story, there is also the other half when it comes to each offspring's genetic make-up. In every parent-and-child pair, precisely one-half the genes of each are *not* shared. Previous observers have focused on the glass half full—the convergent part of genetic identity, ignoring the other, empty half.

Not only that, but biologists—and most psychologists, psychiatrists, and sociologists to an even greater extent—have treated the child as an appendage to the parent, rather than a separate being with its own strengths and weaknesses and, more important, its own agenda, dictated by the fact of one-half genetic *non*overlap. By taking this nonoverlap seriously, new dimensions of parent–offspring noncooperation become apparent and predictable (how stupid of the rest of us not to have thought of this!).

Trivers suggests that we consider a newborn infant, say, an elephant calf, since the biology of parent–offspring conflict is not unique to our own species, and it might be helpful to start off with a bit of distance from *Homo sapiens*. The interests of infant and mother initially coincide: the calf needs various things from its mother, milk in particular. And the mother is prepared, even eager, to meet those needs. In the short term, her hormones as well as her anatomy predispose her to lactate. But then something happens. The infant grows older, larger, and less dependent on mama's milk. At the same time, the mother becomes inclined to discontinue nursing; after all, milk is energetically costly to produce and at some point, the mother will do better in terms of her own fitness if she stops investing in her current child and prepares to put precious resources into another. (In many mammals, lactating females are inhibited from ovulating, so the nourishing of one offspring literally precludes making another.) In itself, this need not lead to conflict—*if* the infant agrees with the mother. Unfortunately, it usually doesn't.

The mother, we must recall, is selected ("naturally") to make the most of *her* fitness, not necessarily that of her offspring. In fact, her only reason for creating that offspring in the first place is as a means of advancing her own evolutionary fitness. From an evolutionary perspective, offspring are parents' way of making more parents. The infant, by the same argument, is ultimately interested in making the most of *its* fitness, not its mother's. More precisely, the infant is only 50 percent related to its mother but 100 percent related to itself. As a result, the infant devalues its mother's interests by a factor of one-half, which the mother reciprocates; that is, mother and infant are each only one-half invested in the other's success, or—no less important—only one-half as averse to the other's costs.

Think of the therapist's cliché, "I can really feel your pain." Infant and mother can each feel only one-half of the other's pain.

The upshot is that after a period in which mother and infant are in agreement about nursing, because both benefit by it, a predictable zone of conflict arises. The mother and infant become locked in a battle of evolutionary wills, with the infant selected to demand more than the mother is selected to give. The infant, after all, gains in proportion as it literally obtains additional nourishment, the only downside being that insofar as continued nursing inhibits the mother's ability to reproduce again, a greedy, hungry infant is deprived of a follow-on sibling. But each of its genes would only have a 50 percent probability of being present in that sibling—assuming they had the same father. If a different father, then only a 25 percent probability of genetic identity with its half-sibling. Either way, the infant can be expected to be more enthusiastic about its own welfare than about the possible future existence of a half or full sibling.

It is misleading, by the way, to think that since any given infant, when he or she eventually reproduces, would have a 50 percent genetic overlap with its own offspring—which equals the 50 percent genetic overlap with a full sibling—said infant should be as interested in the sibling as in itself. Bear in mind that the relevant comparison is between self and sibling, which equals 100 percent versus 50 percent, for a bias of 2:1. Or proceeding down one generation, between self's offspring and sibling's offspring, for a bias of 50 percent versus 25 percent, or 2:1 once again. However you slice it, "self" trumps "sib."

The mother, by contrast, has an equal genetic interest in each of her offspring, whether the one currently being nursed or the next, potential one. The existing infant, however, is one-half as interested in getting an additional sibling as it is in getting as much nourishment as possible, so as to be able to produce its own offspring someday. No wonder weaning conflict is so widespread!

Fortunately, there is light at the end of this tunnel. For the mother, the cost of nursing continues to mount, while for the infant, the benefit of nursing begins to decline. Eventually they coincide: it becomes in the infant's interest for the mother to stop giving so much and to start taking care of herself—or, another way to see it, the infant eventually wants the mother to provide it with siblings, so as to enhance the infant's biological fitness, especially since its need for additional milk has been declining. Eventually, after all, it will need to nourish itself. The mother is only too happy to oblige (in order to maximize her own fitness), and nursing is finally discontinued.

Before then, weaning conflict is rampant. Observe a cat with kittens: the mother encourages most of the early nursing bouts, until the young are around three weeks of age. For the next ten days or so, kittens and mother are equally likely to initiate nursing. By about day thirty, however, it is the kittens who attempt to nurse, while their mother discourages their efforts, to the point that she is likely to get up and walk away.[3]

Something similar even takes place among birds. Large nestlings—big enough to fly, hence known as fledglings—can often be found pursuing their harried parents, importuning them for food. Throughout North America in late spring, it is common to see fledglings quivering their wings and uttering incessant begging calls, while the parents back away, look far into the distance (as though trying to ignore what is in front of them), and often literally take wing, pursued by their nearly grown, but indefatigably demanding offspring. One of the most comical examples is the so-called feeding chase of flightless Adélie penguins, in which the adults waddle about all over the rookery, desperately pursued by rapacious juveniles.

Conflict over weaning, or its equivalent in birds, does not begin to exhaust the potential for parents and offspring to disagree. When a rhesus monkey is just born, the infant stays close to its mother. When the baby wanders off, nearly always it is retrieved by the mother. By age fifteen weeks or so, however, the tables have turned, and it is the young monkey who seeks to initiate contact, only to be increasingly rebuffed by the mother.[4]

The transition from conflict to offspring independence, although rarely smooth, is often accomplished gradually. Maybe this is because parents are selected to minimize the stress felt by their offspring, or because with parent and offspring so intensely engaged in the struggle, neither party can win abruptly and cleanly. A European species of woodland bird, the spotted flycatcher, reveals a nice example of conflict combined with a gradual transition to offspring independence. During the first nine days after hatching, spotted flycatcher parents feed their young regardless of whether these offspring are silent or calling. After ten days or so, however, feeding only takes place when the young call vigorously. Chicks also begin chasing their parents, demanding food, which parents are increasing reluctant to provide. As time goes on, food is *only* given up after a chase, and sometimes not even then. By day eighteen, only 20 percent of feeding chases end in the offspring cadging a meal. Moreover, the actual size of such a meal goes down, despite the fact that larger nestlings presumably need more nourishment, not less.[5]

By this time, the offspring show every sign of being hungry, and they start filling their bellies, at the expense of nearby insects. The outcome—whether directly intended or not—is that growing spotted flycatchers have been gradually induced to earn their own living: initially, they were fed, whatever they did. Then, they had to call. Then they had to chase their parents, ever more determinedly, and for a longer time. In the end, they had no choice but to forage for themselves. The offspring become progressively better at catching their own food while still being provisioned—although in ever smaller amounts—by their parents, who provided a safety net of sorts. But eventually, like it or not (and to look at them, mostly the answer is "not"), the young become independent, and the parents, off the hook.

More generally, these and other conflicts are over what Trivers in an earlier study had termed "parental investment"—the various expenditures of time, energy, resources, and risk that parents make in their efforts to care for and promote the success of their offspring.[6] It is not hard to draw parallels to people. It fact, it is hard not to do so.

WITHIN OUR OWN species, parent–offspring conflict is variable, but probably goes on longer and more intensely than in any other creature. Many a harried parent, struggling to provide for even the most rewarding and undemanding child, will answer the question, "What do you want your child to be?" with an immediate reply, "Self-supporting!"

Teen-agers are often encouraged to work part-time or on weekends, if they want extra spending money, or a car. Then there is the question of college. Some parents pay for it willingly; others only grudgingly. Yet others, not at all. Clearly, as offspring grow older, some sort of transition is reached, although most of the time, children would appreciate more parental investment (albeit less meddling) than parents are inclined to provide. The joke goes as follows:

Son writes to father, after a few months at college—"Dear Dad: No mon, no fun, your son." Father responds—"Dear Son, Too bad, so sad, your dad." I would bet a similar dialog obtains cross-culturally, but this remains to be demonstrated.

The implications of the theory of parent–offspring conflict are profound, with much yet to be explored. Trivers points out, for example, that conflict can be expected not only over the timing of the withdrawal of parental investment but also over the amount provided. How much, for example, is a parent predicted to invest in a given offspring at a given time? Answer: whatever it takes to maximize the donor's fitness, until the parent's cost of such provisioning exceeds his or her benefit. Offspring, however, are expected to see things differently, since each one devalues parental cost by a factor of one-half. The result: conflict once again.

Parents and offspring can also be expected to disagree in predictable ways regarding behavior toward a third party. Consider two siblings. Given the probabilities of genetic relatedness, we expect one sib to help another (be "altruistic") whenever the cost to the altruist is less than one-half the benefit derived by the recipient. That way, the altruist comes out ahead because he or she devalues the sib's payoff by one-half: if that recipient sib gains more than twice what the altruist loses, then in the hard-headed calculus of evolution, the exchange ends up being selfish after all, at the level of the genes involved.

Parents, however can be expected to disagree, to encourage altruism from one sibling to another whenever the recipient's benefit exceeds the donor's cost (without the one-half devaluation), because parents are equally related to each of their offspring and can therefore to be expected to value each of them equally. The result is yet another predicted zone of conflict, with parents anticipated to urge their children to share, and play more nicely—in fact, twice as nicely!—as the children are themselves inclined.

Picture two sibs, Billy and Jane. Let's imagine that Billy wants some of the food Jane is eating. If Billy's need for it is greater than the benefit that Jane derives from it, then their parents would want Jane to share. Not only would Billy benefit from the gift, but so would the parents, so long as Jane's cost in giving it up is less than Billy's gain from receiving it. But Jane can be expected to see things her own way: granted, she is 50 percent related to Billy, but she is 100 percent related to herself. Hence, she is predicted to be more selfish than her parents would prefer.

To this, add the crucial caveat "all other things being equal"—which, of course they never are. It is assumed, for example, that each child is equally likely to be successful, and also that each is equally needy. In other words, they are all comparable in their ability to convert parental investment into fitness: their own, that of their parents, or that of a sibling or anyone else. This assumption is never true in real life, but is essential to any basic model, and—in most cases at least—is unlikely to be misleading.

Insofar as Jane is already well fed, she, too, can benefit by sharing with her brother. But there is a crucial difference between her calculus and that of her parents. Whereas Mom and Dad want sib–sib generosity, sharing, kindness, and so forth any time Billy's gain exceeds Jane's loss, Jane drives a harder bargain, since natural selection would have acted on her ancestors (actually, their genes) to value herself literally twice as much as she values her brother, and only to behave in accord with parental wishes when Billy's payoff is twice or more that of her loss.

The above is just one specific case of a common parent–offspring experience: parents urging their children to play more nicely with each other than each child is inclined. Or urging a child to share when she is predisposed to be selfish. There is a phrase for this whole business: sibling rivalry. Thanks to the above evolutionary insights, we can finally go beyond common usage and Freudian hokum, getting to the bottom—or at least, closer to the bottom—of this peculiar phenomenon.

Part of the power of the evolutionary approach comes from the recognition that such cases are not due to sheer stubbornness or perversity on the part of an unaccount-ably rivalrous or self-centered child; rather, children are inclined to act in response to *their own* interests rather than those of their siblings, their parents, society, or anyone else, for that matter.

Think, as well, about cousins. Genes within an aunt or uncle have a one-quarter probability of being present in a niece or nephew: a one-half overlap between the full-sib parents of that niece or nephew, devalued by another one-half commonal-ity between that individual and its offspring. By contrast, cousins—the offspring of two full sibs—have only a one-eighth probability of genetic overlap. Given that aunts/uncles have twice the probability of sharing genes with their nieces/nephews than cousins have with each other, parents can be expected to urge their offspring to be twice as involved with their cousins—nieces or nephews of the parent—than the off-spring themselves are inclined.

In the United States, at least, it is common for cousins to spend time together. But this is notably the case when the cousins are young, at which time interactions are largely controlled by their parents. As time goes on, however, the family grows older and parents become less influential and eventually die. At this point, no longer subject to parental pressure, cousins typically drift apart, or at least they tend to become less close than in the past, when they were in their parents' orbit and less able to assert their own interests.

How will the various predicted patterns of parent–offspring conflict be resolved? Here, too, Trivers makes some provocative suggestions. He notes, for example, that given the physical mismatch between nursing mother and infant, the latter can hardly be expected to fling the former to the ground and demand more milk than the she is prepared to provide. On the other hand, offspring are not without options, particularly psychological ones. They may well, in fact, have become experts at psychological warfare.

Despite anticipated conflicts, parents and offspring undoubtedly have a fundamental shared interest in the latter's success, and offspring are far better positioned to know their degree of need. As a result, they can be expected to employ numerous psychological tactics such as crying, smiling, and so forth, depending on circumstance and opportunity, to get their way, although they would be ill-advised to grossly overrepresent their neediness, lest parents give up on them entirely. At the same time, while they might want to emphasize their competence and potential, thereby making themselves appear to offer what economists call "return on investment," they don't want to appear too good, lest they be deemed capable of flourishing on their own.

For their part, parents can be expected to distinguish—or at least, attempt to distinguish—proclaimed need from the real thing. They're well advised to refrain from being evolutionary suckers, excessively manipulated by their offspring, yet at the same time, any genetic tendency on the part of parents to be indifferent to their offspring's genuine requirements would be strenuously selected against, since after all, in the long run it is almost as much in the parents' interest to have their offspring succeed as it is in the child's interest to do so. (This is why the time of parent–offspring conflict is sandwiched between two periods of parent–offspring agreement.)

Offspring know, better than anyone, what their real needs are, and how severe they might be. Moreover, since it is in the parents' interest to be attuned to these needs so as to respond appropriately, there is genuine benefit to accurate communication. Unless in the grip of especially intense parent–offspring conflict, parents respond when offspring cry, or when they give other indications of distress. Any genetic tendency by the parents to be indifferent to their offspring's requirements would likely be selected against, and this, in turn, gives the offspring just the opening they need. They can proceed to manipulate—or attempt to manipulate—parental gullibility by sending false signals: pretending to be needier than they really are, so as to get the extra amount of parental investment they desire and that parents are reluctant to provide.

On the other hand, infants would have been selected to exaggerate their competence no less than their neediness. When food is scarce or if they are in poor condition, parents could be tempted to cut their losses and cease investing in costly children, especially if the parents might have an opportunity to breed again some other day. In this case, children would be under evolutionary pressure to send potentially reassuring signals—accurate or not—that say, in effect, "I am healthy and strong. I therefore won't take much from you, and moreover, I will probably provide a good evolutionary return on your investment."

Given the choice, parents would likely favor offspring whose signals of this sort are especially likely to be honest, that could not readily be faked. In short, selection should favor offspring signals that are expensive to produce, since those in poor condition would necessarily be less able to send signals of this sort. In this respect, it is interesting that crying by human infants is quite costly, involving an expenditure of calories that is about 11 percent higher than a quiet baby. In addition, various illnesses and debilitating conditions produce consistent variations in "cry characteristics," variations to which parents respond. Crying, in short—especially on the part of healthy babies—may be a way of proclaiming them to be worthy of parental investment.[7]

Fortunately for us all, children don't only cry and complain. They also smile when they are happy, that is to say, when their needs are met. Not surprisingly, adults find this gratifying, to the point of performing all sorts of peculiar antics in the hope of generating a smile from an infant or young child. Children could accordingly withhold their smiles until their needs have been satisfied. Similarly with sleep. For exhausted, sleep-deprived parents, sometimes the greatest reward their infant can provide is to go to sleep. Almost certainly, the child is not saying to itself, "Don't go to sleep until all of your needs are met," any more than it is plotting, "Hold off on that smile until you get a bottle, or a breast, or a goofy look from Uncle Seymour." But it would make biological sense if that same child were to feel agitated and restless until its (unconsciously) desired outcome is achieved, and to act accordingly.

One extreme of agitation and restlessness is the temper tantrum, in which a child behaves in a way that is not only unpleasant for the parent but also may even threaten to injure the child itself. Picture a child beating its head against the wall, or refusing to eat unless or until it gets its way, in a game of chicken played against its own parents. Temper tantrums, or their equivalent, have even been described for animals. Here is the wildlife biologist George Schaller's description of white pelicans in Yellowstone National Park:

> Young, ten or more days old, often begged vigorously for their food. Usually a young pelican sat very upright in front of its parent, with neck stretched high and wings beating, until it was admitted to the pouch [allowed to feed]. Sometimes, however, a young bird ran to an adult, threw itself on the ground, and beat its wings wildly, all the while swinging its head from side to side. Occasionally the young lay on its side, beat one wing, suddenly jumped up, ran at and pecked several young in the vicinity, driving them away, only to continue begging. It also grabbed, shook and bit its own wing with the bill as it turned its body around and around, growling all the time. In the words of Chapman (1908, pg. 102) the young "acts like a bird demented."[8]

And here is Jane Goodall, recounting the trials and tribulations of parent–offspring interactions among chimpanzees:

> Temper tantrums are a characteristic performance of the infant and young juvenile chimpanzee. The animal screaming loudly either leaps into the air with its arms above its head or hurls itself to the ground, writhing about and often hitting itself against surrounding objects. The first temper tantrum observed in one infant occurred when he was 11 months old. He looked around and was unable to see his mother. With a loud scream he flung himself to the ground and beat at it with his hands, and his mother at once rushed to gather him up. Two infants showed tantrums in connection with weaning and this has been recorded also in infant baboons and langurs . . .. Yerkes (1943), when describing tantrums, comments that he often saw a youngster "in the midst of a tantrum glance furtively at its mother or the caretaker as if to discover whether its action was attracting attention." In captivity, individuals are less prone to indulge in temper tantrums as they grow older, and this was also true of wild chimpanzees.[9]

It seems likely that parent pelicans and chimpanzees are as troubled by their offspring's tantrums as human parents are. Probably, they are also about as likely to give in.

Parents—human as well as non—would be selected to distinguish indications of real need from those that are exaggerated and dishonest. As the folktale advises, it may be dangerous to "cry wolf," in this case to insist on being needy when you are merely greedy. Parents can presumably learn to see through such exaggerations, but they are unlikely to discount them altogether. When in doubt, it may be prudent for parents to err on the side of leniency and generosity, especially if the cost to the parent is low and the outcome might be true offspring distress or even death if the child is really in trouble.

It can also be predicted that older parents would be better at discriminating whether offspring solicitations spring from genuine need or from their involvement in parent–offspring conflict, if only because they are more experienced.

Trivers proposed that the psychological phenomenon of "regression" may have its roots in the parent–offspring tug of war, since younger offspring are usually more in need, less likely to be in conflict with the parents' self-interest, and thus most likely to send signals to which parents respond positively. Hence, masquerading as being younger than one is could conceivably be an effective strategy for cadging additional positive attention and even resources.

Most people find babies and toddlers attractive. Also puppies, kittens, and so forth. "The only thing wrong with a kitten is that / it grows up to be a cat." So wrote Ogden Nash, giving words to a nearly universal human weakness for animals that are young, vulnerable, and thus, cute and particularly endearing. Given that people are especially susceptible to signals that indicate helplessness and extreme youth, it is to be expected that they would have also evolved to send signals that mimic those same traits.

On the other hand, parents, too, have psychological techniques at their disposal. Because of their greater age and experience, they often have something worthwhile beyond sheer resources and protection to transmit to their offspring. Offspring, in turn, are therefore likely to be vulnerable to parents who exaggerate their wisdom, just as parents are vulnerable to offspring who exaggerate their neediness, or even, their competence. It may therefore be significant that the traditional view of parent–offspring relations, which assumes that "father and mother know best," and which has never taken the perspective of offspring very seriously, is one that has been promulgated by adults.

Similarly, we can expect that parents would be likely to represent their teachings, manipulations, and arm-twisting as "for your own good," or even accompanied by protestations that "this hurts me more than it hurts you"—in short, to emphasize the value of their opinions and advice, as well as the degree to which the parental perspective is solely in the best interest of the child.

OF ALL THE new, paradigm-busting insights provided by a gene's-eye view of evolution, it may be that parent–offspring conflict has been the least explored. It has barely been acknowledged by child psychologists, and not at all by psychoanalysts. Yet it may well be that the phenomenon of conscience (or the Freudian construct known as the superego) represents a kind of parental victory, whereby parents induce their offspring to act in their—that is—the parents'—best interests, rather than the child's.[10]

For another example, take infantile (or rather, childhood) sexuality, a mainstay of psychoanalytic theory, yet poorly supported by facts. It, like the human conscience, might owe its existence to parent–offspring conflict. Here's how: when young children display a kind of intense (erotic?) interest in the opposite-sex parent, this could result in greater inclination by that parent to invest in them, not because of a direct, reciprocal urge by the parent in question, but because it is only "human nature" to be flattered, to respond to such interest and attention by providing additional love, attention, resources, and so forth, in return. If so, then selection could favor such behavior by children, as part of their armamentarium in parent–offspring conflict, a way of counteracting what might otherwise be a parental inclination to discontinue investment or, at least, provide less than the child might prefer. At the same time, it would be in the interest of the child not to be too overt about his or her approaches to the opposite-sex parent, if only because this might arouse the ire of the other parent.

If in fact, fathers are attached with special intensity to daughters, and mothers to sons (which is commonly believed, but not necessarily true), the above could be a partial explanation. From an evolutionary perspective, it is unlikely that children actually harbor desires for sexual union with their parents, and far more plausible that children occasionally respond provocatively to a parent in order to garner additional investment from him or her, as one of their tactical moves in the ongoing conflict between parents and offspring.

Here is another example, this one for the developmental psychologists: the phenomenon of the "terrible twos." It is a commonplace that toddlers go through a particular phase during which their behavior is notoriously trying. Is it merely coincidence that this happens at the same time that the parents are likely to be thinking of having another child? What more effective way for a child to reduce the likelihood of having a sibling earlier than it might wish than by announcing—albeit not in so many words—"I'm already a handful; are you sure you're ready to deal with another one quite so soon?"

It may also be noteworthy that the more an infant "wins" a weaning conflict and prolongs lactation, the more likely he or she is to delay ovulation on the part of the mother and, therefore, the less likely it is that the mother will reproduce again. (The technical term is "lactational amenorrhea," and although it is not absolute in human beings, it is nonetheless real.) Given worldwide problems of overpopulation and undernutrition, prolonged lactation would seem a doubly good idea, and reason to root for the youngster in this particular tug-of-war.

There is also something that may be even more terrible than the "terrible twos," at least from the parents' perspective. It is known as "copulatory interference." The title of the scientific paper first reporting this phenomenon pretty much tells it all: "Responses of Chimpanzees to Copulation, with Special Reference to Interference by Immature Individuals."[11] It turns out that among free-living chimpanzees at Jane Goodall's renowned research site in Tanzania, offspring are less than enthusiastic when their mother tries to copulate. The youngster typically interposes him or herself between the mother and the sexually aroused adult male—a potentially risky undertaking—making distracting noises as well as initiating physical contact with both male and female. During a 16-month period, 341 such instances were recorded.

Nor is this unique to chimps. Similar behavior is common among other primates, including macaques, patas monkeys, vervets, and baboons, and is especially frequent when the mother is just beginning to resume sexual activity, around the time that the infant is being weaned. To be sure, copulatory interference could possibly be due to other causes—anxiety that the mother is being hurt, a voyeuristic interest in adult sexual behavior, and so forth—but it also makes perfect sense in terms of parent–offspring conflict, if such behavior can effectively delay the arrival of a sibling until the infant has received all the parental investment that it can get.

One of Freud's more celebrated cases was known as the "Wolf Man," not because the patient turned lupine during full moons, but because when very young he had apparently witnessed his parents having sexual intercourse, in the "doggie" position. Such "primal scenes" have long been considered emotionally shattering for young observers. There may be some truth to this, although it seems more likely that a toddler would find such occasions confusing than traumatizing.

The real trauma for an inadvertently peeping-Tom child might be that parental intercourse signals the possible impending arrival of a sibling competitor. Hence, we might expect an understandable ambivalence—on the child's part—about whether mom and dad are doing the right thing. (Bear in mind that children need not consciously make this connection, any more than they need to understand the physiology of digestion and metabolism in order to get hungry when they need nutrition.) The good news is that people, unlike chimpanzees, are sometimes fortunate enough to leave their young children with a babysitter and get away for a weekend. Or at least they can lock the door.

Yet another unnerving possibility has turned up: maybe parent–offspring conflict begins even earlier, *before* birth.[12] Thus, some of the complications that bedevil pregnancy may be due to the fact that mother and fetus are not reading from the same genetic script. To be sure, pregnancy is a cooperative endeavor, with the mother seeking to reproduce, and the fetus to be produced. But as the evolutionary biologist David Haig has pointed out, it is consistent with parent–offspring conflict to expect that the fetus will seek more maternal resources (oxygen and nutrients) than the mother will be inclined to provide.

High blood pressure and preeclampsia, for example—two frequent complications of pregnancy, suffered by the mother—are mostly caused by fetal hormones that attempt to increase blood flow to the unborn child at the expense of blood flow to the mother's body. Blood sugar is another possible battleground. During the last trimester of pregnancy, the placenta (a product of the fetus, not the mother), secretes a hormone that interferes with the effect of the mother's insulin. The mother, at the same time, is producing more and more insulin. Why this particular tug-of-war? Her insulin suppresses maternal blood sugar levels, so by making insulin, the mother can be seen as attempting to restrict the amount of sugar—her nutrient—that is made available to the fetus, who in turn responds by trying to get as much as possible.

In most cases, things work out all right. But sometimes, the fetus wins and the result is gestational diabetes. Perhaps it isn't a coincidence, by the way, that the gene responsible for producing the fetus's anti-insulin chemical, human placental lactogenic hormone, is provided by the *father*—who, in evolutionary terms, can be seen as on the fetus's side rather than the mother's.

Evolutionary biologists familiar with the theory of parent–offspring conflict are not surprised that the dominant ideology—both in social science and in normal family life—tends to privilege adult wisdom and good intentions over juvenile "intransigence" and "ignorance." A truly biological perspective suggests, however, that the interests of neither parents nor offspring should be considered determinative; rather, they are interactive, dynamically interpenetrating, and, on occasion, diametrically opposed. In such cases, the likely results are evolutionary arms races, tugs-of-war, and various manifestations of outright conflict, often with no obvious victor. For beleaguered parents and their offspring alike, this may not constitute surprising news, but for supposed experts whose "expertise" may well have been based on a thoroughly inaccurate and nonevolutionary view of their subjects, the theory of parent–offspring conflict promises—or threatens—to generate a new and (r)evolutionary paradigm.

Old paradigm: Parents and offspring are united in their interests, albeit sometimes at odds for other reasons.

New paradigm: Parents and offspring have genuine, predictable, biologically mediated areas of conflict.

NOTES

1. Some of the material in this chapter was modified and repurposed from an article that I wrote for the *Chronicle of Higher Education* (http://www.chronicle.com/article/Parent-Child-Conflict-Its-in/130387/), and from my book *Revolutionary Biology: The New, Gene-Centered View of Life* (New Brunswick, NJ: Transaction Publishers, 2003).
2. R. L. Trivers, "The Evolution of Reciprocal Altruism," *Quarterly Review of Biology* 46, no. 1 (1971): 35–57.
3. T. C. Schneirla, J. S. Rosenblatt, and E. Tobach, "Maternal Behavior in the Cat," in *Maternal Behavior in Mammals,* ed. H. Reingold (New York: Wiley, 1963), 122–168.
4. R. A. Hinde, "Mother-Infant Separation and the Nature of Inter-Individual Relationships: Experiments with Rhesus Monkeys," *Proceedings of the Royal Society of London. Series B, Biological Sciences* 196, no. 1122 (1977): 29–50.
5. N.B., Davies, "Parental Care and the Transition to Independent Feeding in the Young Spotted Flycatcher (*Muscicapa striata*)," *Behaviour* 59, no. 3 (1976): 280–294.
6. T. Robert, "Parental Investment and Sexual Selection," in *Sexual Selection and the Descent of Man* (New York: Aldine de Gruyter, 1972), 136–179.
7. F. B. Furlow, "Human Neonatal Cry Quality as an Honest Signal of Fitness," *Evolution and Human Behavior* 18, no. 3 (1997): 175–193.
8. G. B. Schaller, "Breeding Behavior of the White Pelican at Yellowstone Lake, Wyoming," *The Condor* 66, no. 1 (1964): 3–23.
9. Jane van Lawick-Goodall, "The Behaviour of Free-Living Chimpanzees in the Gombe Stream Reserve," *Animal Behaviour Monographs* 1 (1968): 161IN1–311IN12.
10. E. Voland and R. Voland, "Parent-Offspring Conflict, the Extended Phenotype, and the Evolution of Conscience," *Journal of Social and Evolutionary Systems* 18, no. 4 (1995): 397–412.
11. C. E. Tutin, "Responses of Chimpanzees to Copulation, with Special Reference to Interference by Immature Individuals," *Animal Behaviour* 27 (1979): 845–854.
12. D. Haig, "Genetic Conflicts in Human Pregnancy," *Quarterly Review of Biology* 68, no. 4 (1993): 495–532.

11

True or False?

ONE OF THE most provocative insights of today's evolutionary biology is that animal communication isn't as simple—or as honest—as had been thought. And more provocatively yet, the same is almost certainly true of people, too.

Evolutionary biologists sometimes appear to be cynics, focusing on the self-serving motivations behind even prosocial behaviors, whether of animals or people. Thus, "altruism" is seen as resulting from the impact of genes that are basically selfish. Courtship is viewed through the lens of competitive male-versus-female strategies. As it happens, biologists such as myself probably *are* cynics, and for good reason: the evolutionary process rewards behavior that benefits whoever does the behaving (actually, his or her genes), not necessarily the recipient.

This is especially the case when it comes to communication, something that nonbiologists typically see as a benevolent, shared effort, whose goal is to convey information as accurately as possible. But is there truth behind this cheery, positive perspective?

Let's begin at the beginning: what *is* communication? It won't do in this case to quote Supreme Court Justice Potter Stewart, who, speaking of pornography, famously observed that he may not be able to define it, but he knows it when he sees it—because, in fact, people often do *not* know communication when they see it. Or rather, they are often deceived about it. And that is precisely the point.

The traditional understanding of "communication" is well captured by its Latin origin: the verb *communicare* means "to share," and those who communicate are members of the same "sharing community." For a less prosocial and more stripped-down scientific definition, communication is simply the transfer of information from a sender to a receiver. It is widely assumed that during this transfer, the sender benefits by sending and the receiver by receiving. In the early days of so-called classical ethology, as pioneered by the biologists Niko Tinbergen and Konrad Lorenz, this optimistic and naive view of communication held sway. Sender and receiver were thought to be "on the same page," with senders indicating, for instance, their internal state (aggressive, defensive, sexually aroused, etc.) by various postures and behavior, and receivers concerned only with decoding the messages as accurately as possible.

What gets in the way, according to this perspective, is mostly the physical space between different individuals, plain old distance that must be crossed by all signals—whether visual, auditory, chemical, and so forth—and that accordingly introduces unavoidable confusion and error. There is also the simple fact that sender and receiver are distinct creatures, each with its own sense organs and nervous systems, which introduces yet more opportunities for miscommunication.

In this dance of mutual benefit, the undisputed champion has been a dance itself: the celebrated "dance of the bees," by which a foraging honeybee, having discovered a good food source, communicates its location—with remarkable accuracy—to others within a darkened hive. (This research was conducted by Karl von Frisch, who joined Lorenz and Tinbergen in being awarded a Nobel Prize in 1973.)

There is no reason to suspect a dancing scout bee of dishonesty. No biological payoff has yet been identified whereby a bee, having discovered a bunch of yummy flowers, would misrepresent its location to her sisters. But as evolutionary theorists eventually realized, other cases are likely to be much darker. For most creatures other than bees, communication is no bed of roses. The problem basically is that senders do not necessarily have any commitment to conveying the truth as such. Instead, they—or rather, their genes—are simply interested in maximizing *their own* evolutionary success. And they can often achieve this by influencing the behavior of someone else: in such cases, whatever the recipient is receiving, it isn't necessarily benevolence. Or truth.

We are accustomed to deceit in animal communication when it operates between different species. That's what camouflage, for example, is all about: a ptarmigan, white against the winter snow, says, "I'm not here." A stick insect says, "I'm a stick, not an insect." Other potential prey items claim to be a leaf or some bird droppings. Bright yellow circles inscribed on the wings of a moth, which—when they suddenly pop open—and deceive a would-be moth-eater by revealing what seems to be the eyes of an owl that preys on whatever might otherwise be trying to prey on the moth.

These examples should italicize the important fact that when it comes to the underlying biology of communication, neither honesty nor dishonesty implies anything about consciousness or intentionality. A moth whose wings possess owl-mimicking eyespots has no need to know that this is the case or that deception is under way. The overwhelming likelihood, in fact, is that he or she has no such knowledge. Many flowers resemble bees or wasps, thereby inducing a foraging insect to copulate with them, in the process pollinating the next flower to be sexually accosted. It is highly unlikely that the plants in question are aware of the deception; rather, flowers that did a good job seducing insects left more descendants than others that were less sexually appealing.

But what about communication within the same species?

Once more, intent to deceive—or, for that matter, to be honest and straightforward—is simply not a necessary or even a likely part of the evolutionary process, at least when it comes to creatures with small brains or (in the case of plants) no brains at all. But when it comes to smart creatures, presumably including *Homo sapiens*, intentionality can certainly be part of the mix. Even then, however, mindfulness doesn't always lead to mutual understanding. And even the smartest people don't always know what they are communicating, for reasons to be explored shortly.

One possibility is that "What we've got here is [a] failure to communicate," as a brutal prison warden states in the movie *Cool Hand Luke*, when "Luke" (played by Paul Newman) has been recaptured following an unsuccessful escape attempt, and just after he has been viciously beaten. More likely, what we've got—not just "here" or "there" in one notable movie, but all over the place—is an effort, presumably quite intentional, by the sender to manipulate the receiver. And in the case of the warden, fury when the attempt failed.

In some cases, such as those marvelous dancing bees, no deception is expected because genetic interests are closely shared. But bees (ditto for ants and wasps) are unusual in that their peculiar chromosomal arrangement results in an exceptionally close overlap among the DNA of individual workers. Most living things—including human beings—are not biologically concerned with the success or failure of other, unrelated individuals. Add to this the fact that individuals and their genes are selected to maximize *their own* benefit, not that of others, and we can expect that their interest in honesty and accuracy is rather limited. What does interest them is making the most of their evolutionary situation, which typically includes using as little energy and running as little risk as possible—all the while accomplishing whatever is to their genetic benefit.

"When an animal seeks to manipulate an inanimate object," write the biologists Richard Dawkins and John Krebs,[1]

it has only one recourse—physical power. A dung beetle can move a ball of dung only by forcibly pushing it. But when the object it seeks to manipulate is itself another live animal, there is an alternative way. It can exploit the senses and muscles of the animal it is trying to control, sense organs and behavioural machinery which are themselves designed to preserve the genes of that other animal. A male cricket does not physically roll a female along the ground and into his burrow. He sits and sings, and the female comes to him under her own power.

In short, *communication may be indistinguishable from manipulation.* It is a stunning, cynical, and distressing insight—especially for those of us who make our living by communicating!

At a concert I attended roughly half a century ago, satirist Tom Lehrer observed "If a person can't communicate, the very least he can do is to shut up." Fair enough. But Mr. Lehrer was speaking as a brilliant satirist. What about the fact that no one—except perhaps the most committed hermit—lives without communicating, at least on occasion? Not only is it difficult to shut up, but the reason most of us find it difficult is that we have "skin in the game," payoffs either obtained or foregone as a consequence of whether we succeed in communicating. Which is to say, whether we succeed in getting something from the transaction, which is often less an exchange than a competition.

Maybe the issue is more whether people (and certain animals) can or cannot succeed in communicating what they want—that is, whether they are able to manipulate their "partners" in this particular and surprisingly complex interaction. And an important contributor to the success or failure of would-be communicators—which is to say, all of us—is the extent to which the recipients are able to decode, evaluate, and often, to resist, the messages being sent.

It isn't easy, at least in part because human beings may well have been primed by natural selection to believe what we are told, or what we perceive, especially if the information in question relates to matters of well-being or survival. Imagine that you are an early protohuman striding through the African savannah. You hear a faint whispering sound; it might be the wind . . . or the exhalation of a crouching leopard. In neither case is the sound intended for your ears, but it is certainly information, with a sender and a receiver. If you, the receiver, conclude that the sender is merely the wind whereas it is actually a dangerous predator, there is a good chance that you won't make this mistake a second time, and your error-prone genes will be that much less likely to be promoted into future generations. On the other hand, making the opposite error—mistaking the wind for a leopard—might cause some wasted time and unnecessary anxiety, but the outcome will be much less dire.

More generally, the outcome of this kind of situation is likely to be selection for credulousness, a tendency to be trusting and to take information seriously, even to the point of gullibility. On the other hand, insofar as many of the signals we receive are generated by other biological entities with a particular interest in manipulating us, inducing us to do things in their interest (the senders'), selection would call a halt to outrageous naivety, just as it would operate against being so jumpy and distrustful that our ancestors couldn't go for a stroll, gather roots, or hunt gazelles without panicking at every natural but inconsequential sound, sight, or smell. Ideally, we are not only open to new and potentially important information ("there's a leopard nearby") but also able to disregard signals that are irrelevant (wind in the grass), and also to identify misrepresentation (as when a perfectly edible berry resembles a poisonous one, or when a gazelle's coloration helps it blend into its surroundings). Along with the capacity to misrepresent, evolution has bequeathed us a parallel capability to decode the misrepresentations of others. The result is an unending arms race, with senders sending and receivers assessing.

IN SHAKESPEARE'S PLAY *Henry IV, Part I*, the mystical, charismatic Welshman Owen Glendower boasts, "I can call spirits from the vasty deep," to which Hotspur responds as befits a blunt Englishman, "Why, so can I or so can any man; But will they come when you do call for them?" Under what conditions would those spirits emerge from the vasty deep? Hint: not simply because the glowering Glendower calls. Rather, the view from biology suggests that they should only come when it is in *their* interest—not Glendower's—to do so. After all, Glendower might want to harpoon or saddle them, or perhaps subject them to unspeakable tortures for his own perverse gratification. On the other hand, maybe those same spirits will profit by being called: perhaps the egotistical Welshman has good news, something of value to share, or useful information to impart—useful to those submerged spirits, that is.

All receivers should accordingly be selected to discriminate self-serving from beneficial "calls" emanating from the likes of Glendower (which is to say, from anyone). And insofar as he gains by inducing the spirits to respond to his efforts, Glendower, in turn, should be selected to send messages that would appeal to the apparent self-interest of the recipients, whereas in reality, they more likely contribute to his own. In short, given the fundamental self-interest of the evolutionary process, communication may well be *dishonest.*

Consider, for example, the case of a male wishing to convince a female that he is a suitable mate. Since males are typically capable of copulating many times, it is often in their interest to persuade females to choose them over any competitors. This puts females in a position of having to discriminate among various suitors, who, in turn, are therefore likely to exaggerate and if necessary, misrepresent their quality as a mate, essentially sending this message: "Choose me, I'm the healthiest (or the strongest, the one with the best genes, or the best provider, and so forth)." Insofar as these representations depart from the truth, they are the animal equivalent of lying. Females, in turn, are selected to see through the false advertisement, leading once again to a mutual arms race: greater deceit by the senders leads to enhanced ability to discriminate on the part of the receivers, which in turn leads to yet more devious deception, countered by yet more sophisticated discrimination, and so forth.

The resulting dance of deception isn't limited to fictional tales of mystical Welshmen or to the much-touted battle of the sexes. When different troops of vervet monkeys meet, someone (typically from among the low-ranking males) may well give an alarm call, of the sort normally reserved for when a leopard is on the scene.[2] The result? Everyone heads for cover, and a violent intergroup clash—in which low-ranking males would typically fare poorly—is avoided. It is possible that such calls during intergroup encounters are simple mistakes. But this is unlikely, since vervets are remarkably sophisticated when it comes to alarm calling (for example, they employ three distinct calls to warn of aerial predators, ground predators such as leopards, and snakes). Moreover, it stretches credulity that only low-ranking males—those who profit from the act—should consistently be the ones making the same errors.

Here is another example of manipulative alarm-calling, once again by animals. The paternity of male barn swallows is threatened by the prospect that females will engage in "extra-pair copulations" with other males. Alarm calls, not surprisingly, break up any such trysts, since the danger posed by a predator is greater than the temptation of adultery. This observation led to a field study in which female barn swallows were chased from their nests while the male was out foraging. Upon his return, finding "his" female absent, male barn swallows gave alarm calls, which would likely disrupt any ongoing extra-pair copulation. Significantly, such manipulative signaling only took place when females were fertile; moreover, it was only performed by swallow species that breed in colonies and not by those that are essentially solitary—whose males are not at risk of being cuckolded.[3]

A bit poignant, this, carrying a nontrivial hint of Rousseauian aroma: the solitary critters are the ones most inclined to communicate honestly. The more social, the greater the tendency to manipulate (and the greater the need to defend one's self against being manipulated). Need I emphasize that *Homo sapiens* is also *Homo socialis*? When candidate Jimmy Carter announced, "I'll never lie to you"—especially notable in the aftermath of Richard Nixon's persistent dissembling—he doubtless gained some support.

One research study found that on average, people tell six to eight lies per day.[4] Hence, our inclination to engage in misleading communication is not limited to such headline cases as perjury, advertising, sexual deception, and politics. It is fundamental to the not-so-simple biological reality that each person is genetically distinct—except for identical twins—and therefore endowed with a kind of existential, evolutionary loneliness that predisposes most of us to manipulate each other, if not most of the time, then more often than even the most honest among us would care to acknowledge. There have been exceptions, of course: truth tellers who are honest about their own lies. In his book—appropriately titled *Human, All Too Human*—Friedrich Nietzsche thought back on how many lies he had invented in order to write as he had done in his younger days, noting "How much falseness I still *require* so that I may keep permitting myself the luxury of *my* truthfulness? Enough, I am still alive; and life has not been devised by morality: it *wants* deception, it *lives* on deception."[5]

A KEY POINT—of this chapter, and, sadly, of life itself—is that mendacity is built into most biological systems, such that the fault lies not so much in advertisers, political leaders, or even in ourselves, but in natural selection: the truth, the whole truth, and nothing but the truth? No way, says natural selection (or so it appears). When it comes to our own species, not surprisingly, the situation is complex but not qualitatively different from that found among other animals. Sometimes we tell the truth—"What time is it?," "Do you have change for a dollar?"—but often not: "Does this make me look fat?," "The dog ate my homework," "This perfume will make you sexually irresistible," "The check is in the mail." A key point is that dishonesty permeates human communication, and not just in the more obvious modalities such as advertising, politics, and barroom braggadocio. Nor can apparent deceit simply be attributed to a "failure to communicate." Often, the only failure involved is a failure to bamboozle the recipient, which itself may be due to insufficient skill on the part of the purported communicator, or especially adroit decoding by the recipient.

The more that the interests of sender and receiver overlap, the more will honesty prevail, and vice versa. But such interests aren't black or white, limited to either complete overlap or irreconcilable conflict. Even the same two individuals will typically have commonalities of interest in some dimensions, along with others that vary from indifferent to downright antagonistic. Two Brazilians, for example, will likely share a predisposition for celebrating Mardi Gras, while at the same time they may work for competing companies, root for different soccer teams, and be courting the same woman.

Along with the temptation for communication to spiral into manipulation—with sender manipulating receiver, and receiver benefiting if she can see through such efforts—there are also opportunities for the receiver to respond by becoming a sender, whereupon she can enjoy a payoff by manipulating the initial, would-be manipulator. Psychologists, for example, have been much taken of late by the phenomenon of theory of mind, sometimes summarized as "mind reading." The idea is that especially in a highly social species, individuals will benefit if they can anticipate another's behavior, often achieved by acquiring a mental picture of what the other individual is likely to do next. To do this, it helps to become familiar with the behavior of one's colleagues: for a dog to know, for example, that when another dog growls and shows its teeth it is liable to bite.

But this is just the beginning. Such familiarity could then lead to a situation in which a growling dog might be well advised (read here: favored by natural selection) to sound menacing and show its teeth even if it has no intent to follow through. Fully six centuries ago, Machiavelli pointed out in *The Prince* that "men judge generally more by the eye than by the hand, because it belongs to everybody to see you, too few to come in touch with you. Everyone sees what you appear to be, few really know what you are."

There is a Spanish saying, notable not only for its poetry but also for its wry validity: "Caras vemos, corazones no sabemos" (*Faces we see, hearts we do not know*). Yet despite the limitations on true, honest, heart-revelatory communication, human beings in particular are overwhelmingly predisposed to keep trying—or maybe to keep manipulating, or to see through others' attempts at manipulating ourselves, or maybe to achieve something like Captain Ahab's yearning to "strike through the mask" and touch the reality underneath. We are, among other things, *Homo communicans*. So we keep at it, at least in part because for human beings in particular, communicating is almost as natural as breathing.

The amount of time and energy people spend communicating, or trying to do so, is stupendous and perhaps literally immeasurable. And yet, the amount of attention directed at understanding the phenomenon itself is downright miniscule. This has been changing of late, with the revolution in computer-assisted communication: e-mail, the Internet, cell phones, texting, Twitter, YouTube, and growing interest in understanding how these new modalities impact our ancient predilections. Along with these electronic "assists," there have arisen new opportunities for manipulation, new threats to privacy, new openings to send, receive, decode, obscure, encrypt, translate, overcome, and influence the almost frantic tsunami of signals that surround us and that we create for ourselves, often with little insight into what we are actually doing.

Only recently, for example, has the biology of eavesdropping been explored, situations in which interactions between individuals A and B are observed or overheard by individual C and others.[6] Moreover, it increasingly appears that those A–B interactions often take place primarily for "the benefit" of the eavesdroppers—which is to say as a way of influencing the eventual behavior of the audience, individuals C and D, toward A or B. In other cases—and this speaks directly to electronic eavesdropping of the sort recently revealed to be the bread-and-butter of the National Security Agency—communications between A and B are biased toward secrecy and privacy.

Whether predisposed toward "public broadcasting" or secrecy maximizing, people are obsessed with communicating, often publicly proclaiming their honesty even while engaging in private deceit. We are definitely not unique when it comes to the use of lies, but human beings are likely in a class unto ourselves when it comes to the degree that we deceive *ourselves*. Thus, the best liars are those who do not even know that they are lying, so that various unconscious signals—blushing, rapid blinking, avoiding another's gaze—do not give away the enterprise.

"Pretending is the brain's work," writes E. L. Doctorow in his novel *Andrew's Brain*. "It's what it does. The brain can even pretend not to be itself."[7] Or as the wolf might well have said to Little Red Riding Hood, "All the better to deceive you with, my dear."

As already described, there are many cases—even among animals—in which honesty seems likely to be the worst policy. Consider two lizards confronting each other over a mutually desired resource such as a nest site, a mate, or a morsel of food. They communicate via threat displays, each seeking to induce the other to back down, thereby avoiding a potentially damaging battle. Success means gaining the resource without a fight, but whoever retreats gets nothing. Under such circumstances, it should pay each contestant not only to communicate its size, strength, and determination, but, if anything, to exaggerate these traits, if by doing so it is more likely to come out ahead. Or take a gazelle, being eyed by a cheetah: since a predator is less likely to waste time and energy chasing prey that it cannot catch, it might well be in the interest of potential prey to communicate to a would-be predator that the latter has been seen and has therefore lost the benefit of surprise. Better yet if the gazelle could make it clear that she is so quick, agile, and healthy that pursuit would be unavailing. But what is to stop the gazelle from sending a dishonest message, seeking to indicate that it is quicker, more agile, and healthier than it really is? (In which case cheetahs would no longer take gazelle messages seriously—assuming that they ever did.)

A compelling answer has been proposed by the Israeli zoologist Amotz Zahavi, who suggests that indeed, communication is seriously bedeviled by the temptation to cheat and to send dishonest signals.[8] As a result, argues Zahavi, receivers are likely to pay special attention to messages that are inherently protected against cheating and are necessarily honest because they are expensive and difficult if not impossible to fake. Hence, gazelles that might otherwise be stalked by cheetahs engage in "stotting," a peculiar, high, stiff-legged jump that requires quickness, agility, and overall health. Sickly gazelles can't stot. It would also be a bad idea for them to try, since their failure would italicize their incapacity and thus, their vulnerability. By the same token, females of many species insist that courting males actually present them with genuine prey items, whose nutritive value cannot be faked, just as a male's quality as hunter, scavenger, or provisioner is also guaranteed to be real if he actually has a nuptial meal to offer.

More controversially, Zahavi's "handicap principle" suggests that the choosier sex (generally females) will be selected to prefer potential mates whose courtship proceeds despite the fact that they are operating under a handicap, which serves as a guarantee of quality and thus, of honest communication. Under this view, for example, females prefer to mate with males sporting gaudy secondary sexual characteristics (bright feathers, colorful wattles, elaborate and expensive courtship shenanigans) precisely because these traits, being significant handicaps, are difficult to fake. A parasite-ridden, genetically challenged, or metabolically stressed suitor would be less able to manufacture the elaborate tail of a successful peacock, "boom" like an impressive sage grouse, drive a Mercedes, or be able to swim the Hellespont.

The result, in some cases at least, is a paradoxical tendency for much communication to be at least somewhat honest, after all. And this is a very useful conclusion, since right now I am trying to communicate with you! Besides, I wouldn't try to manipulate you—would I?

Old paradigm: Communication is assumed to be honest, providing mostly truthful information.

New paradigm: It is at least as likely to be dishonest, or in any event, an effort by the sender to manipulate the receiver—for the sender's benefit.

NOTES

1. R. Dawkins and J. R. Krebs, "Animal Signals: Information or Manipulation," *Behavioural Ecology: An Evolutionary Approach* 2 (1978): 282–309.
2. R. M. Seyfarth, D. L. Cheney, and P. Marler, "Vervet Monkey Alarm Calls: Semantic Communication in a Free-Ranging Primate," *Animal Behaviour* 28, no. 4 (1980): 1070–1094; D. L. Cheney and R. M. Seyfarth, "Vervet Monkey Alarm Calls: Manipulation through Shared Information?" *Behaviour* 94, no. 1 (1985): 150–166.
3. W. A. Searcy and S. Nowicki, *The Evolution of Animal Communication: Reliability and Deception in Signaling Systems* (Princeton, NJ: Princeton University Press, 2005).
4. B. M. DePaulo et al., "Lying in Everyday Life," *Journal of Personality and Social Psychology* 70, no. 5 (1996): 979.
5. Friedrich Nietzsche, *Human, All Too Human* (Stanford, CA: Stanford University Press, 2012).
6. E.g., T. M. Peake, "Eavesdropping in Communication Networks," in *Animal Communication Networks* (Cambridge: Cambridge University Press, 2005).
7. E. L. Doctorow, *Andrew's Brain* (New York: Random House, 2014).
8. Amotz Zahavi and Avishag Zahavi, *The Handicap Principle* (New York: Oxford University Press, 1999).

12

The Myth of Monogamy

MOST PEOPLE PROBABLY agree that it would be nice if human beings were naturally monogamous. In fact, most biologists, anthropologists, and psychologists would likely agree as well. There are many reasons for this preference, going beyond Judeo-Christian moralistic expectation as well as the mandate of Western legal systems. The sad reality, however, is that we aren't. (For anyone especially forlorn about this aspect of humanity's nature, don't despair; this chapter ends rather optimistically.)

But first, a quick review of the evidence that polygamy is the default human system.

For starters, we look at some of the evidence for polygyny, commonly identified as "harem formation," in which one man mates with multiple women. Then we take quick look at the other side of polygamy, namely polyandry, in which one woman mates with multiple men. "Sexual dimorphism" is a revealing biological fact, highly correlated with polygyny. It shouldn't be controversial to note that men and women are anatomically different—beyond the obvious distinction of beards versus breasts, and penises versus vaginas. Although some men are shorter, lighter, and less muscular than some women, most men are taller, heavier, and have more muscle mass than most women. This difference could be due to many factors, such as selection among prehistoric males for warding off predators, for success in hunting large game, for victory in warlike competition with members of other, competing bands, or perhaps merely for carrying heavy loads or digging ditches. But there is substantial reason to think that such sexual dimorphism is largely due to male–male competition for access to females.

The strongest evidence here is comparative. Looking at a sample of closely related species (among the New World monkeys, Old World monkeys, great apes, deer and their relatives, seals, birds, reptiles, and so forth), the pattern is clear: the greater the degree of polygyny (the average number of females sexually monopolized by one male), the greater the degree of sexual dimorphism. Among seals, for example, elephant seals are the most polygynous, with a successful harem-master sometimes accumulating as many as forty females under his sexual suzerainty. Unsuccessful bull elephant seals end up being resentful bachelors, and because there are equal numbers of males and females, for every harem-master—who gets to father lots of offspring—there can be dozens of reproductive failures. Not surprisingly, bull elephant seals engage in colossal battles with each other; equally unsurprising is the fact that bull elephant seals are nearly elephantine, weighing four to eight times more than the cows, since the payoff to success is to project copies of one's genotype into future generations, and those genes that contribute to building a large and imposing body are the ones thus projected. Other seal species that are less flagrantly polygynous are correspondingly less sexually dimorphic, and those few species that are essentially monogamous are essentially sexually monomorphic—showing no substantial male–female differences.

A similar correlation has been found pretty much everywhere it has been sought, notably including our closest relatives, the great apes. Thus, gorillas are quite polygynous, with large, imposing silverback males monopolizing up to six adult females, with younger blackback males and smaller silverbacks being reproductively excluded. Gorillas are also highly dimorphic, with males outweighing females by two- or threefold. Among chimpanzees, which engage in something of a sexual free-for-all, but in which dominant males nonetheless obtain more than their share of the matings, adult males are roughly 1.5 to 1.7 times larger than females, and gibbons, whose mating system is mostly unimale, sexual dimorphism is mostly nonexistent. That particular great ape species known as *Homo sapiens* reveals a male/female body mass ratio of roughly 1.2, suggesting that our ancestors were mildly polygynous, albeit not wildly so.[1]

The word "dimorphism" derives from "two bodies," an adaptive outcome that would have been substantially less adaptive if our ancestral twofold bodies hadn't been outfitted with appropriately twofold behaviors. And so they were. Behavioral dimorphism is no less apparent than its anatomical counterpart. It would not benefit our large, comparatively muscular male ancestors if they were not predisposed to make use of those bodies, when, literally, push came to shove. The prediction therefore arises that insofar as dimorphism—behavioral as well as physical—is driven by evolutionary striving for polygynous success, those comparatively large males should be comparatively bellicose, too. Sure enough, they are.

Looking across different species, once again, we find that degree of behavioral dimorphism is predictably consistent with degree of polygyny. Thus, not only does sexual dimorphism involve males being typically larger than females but also, in proportion as a species is polygynous, males are consistently outfitted with especially well developed horns or antlers, impressive fangs, extra-long claws, and so forth, along with an inclination to use them. Human beings, to be sure, are notable for lacking built-in anatomical weaponry, perhaps because early in our evolutionary history we developed the habit of picking up and using weapons and other tools. But the data are also overwhelming that not only are men more aggressive and violent than women—cross-culturally—but also significant differences arise very early in childhood.[2]

Next in line: sexual bimaturism. This refers to the widespread but often overlooked phenomenon whereby the two sexes differ in the average age at which individuals become sexually mature. Among mammals, females have by far the hardest job when it comes to producing offspring. They must carry the fetus, nourish it throughout pregnancy, and go through the rigors of giving birth, and then the most demanding part begins: lactation requires substantially more calories than does pregnancy. Males, by contrast, need only produce a few drops of sperm-infused semen, a metabolic investment equivalent to eating a couple of potato chips. Given this investment gap, one might expect males to start breeding while still quite young, and females to delay until they are big enough, old enough, and possibly also experienced enough to handle the demands of reproduction. In polygynous species, however, the situation is precisely reversed, with females becoming sexually mature at an age when males are still functional adolescents—which is to say, not functionally reproductive. In our own species, this dichotomy is especially noticeable among early teenagers, when the girls are blossoming into womanhood while the boys are still obviously boys: socially and sexually less mature, and often several inches shorter than their female classmates.

Polygyny resolves this paradox. Although the physiological demands of breeding are heavier on females than on males, in a polygynous species male–male competition generates a heavy load of social and behavioral competition, which falls most intensely on the sex doing the most vigorous competing. Hence, in every polygynous species thus far studied, males delay their maturation until they are up to the task of holding their own in the often rigorous—and potentially even lethal—arena of sociosexual competition. Among monogamous species, there is no sexual bimaturism. Here, therefore, is additional evidence that human beings are not naturally monogamous. There is yet more.

Looking at the situation among traditional societies when they were first contacted by Western traders, missionaries, explorers, and colonialists, only about 16 percent were officially monogamous. Of the rest, fewer than 1 percent were polyandrous ("reversed" harems, in which one woman mates with multiple men), and 83 percent were avowedly polygynous. Even among these latter groups, it is likely that the majority of men had only one wife—but not because they didn't try to obtain more. Monogamy (or worse yet, bachelorhood) was a default setting for those men unable to garner a harem. The male behavioral compass was nonetheless set at polygyny when possible.

Looking at DNA retrieved from midrange human fossil material, the evidence for primordial polygyny is further confirmed. The Y chromosomes—inherited only by males, from their fathers—turn out to show substantially less genetic diversity than does the mitochondrial DNA, which is inherited only by females, from their mothers.[3] In other words, and as illogical as it might seem, we have fewer male ancestors than female. Those numbers would be equal only if we are descended from strictly monogamous stock; instead, the DNA evidence testifies clearly to the fact that a relatively small number of our great-great-etc. grandfathers mated with a relatively large number of our great-great-etc. grandmothers. In other words, as the researchers concluded, "over much of human prehistory, polygyny was the rule rather than the exception."

There is even more evidence for our polygynous heritage. For example, men show a persistent cross-cultural preference for a greater variety of sexual partners than do women, a difference that shows up not only in surveys but also in physiology, specifically the well-known fact that a new copulatory partner tends to revive an otherwise flagging male sexual response. This phenomenon, known as the Coolidge effect after a possibly apocryphal story,[4] is found in polygynous species generally, and is less consistent among monogamous ones.

OUR DISCUSSION SO far has considered only men and their harem-based polygyny. What about women?

Here the evidence is less clear-cut, but almost certainly no less real, the bottom line being that just as human beings are naturally polygynous, they are also naturally polyandrous. For one thing, although women have traditionally been thought (at least in the prudish West) to evince the inverse of "natural" male horniness, the reality is that women, too, can be sexual adventurers in their own right. (To a remarkable extent, men—including male scientists—have long been remarkably obtuse when it comes to taking a clear-eyed look at female sexuality.) Such obtuseness has been encouraged by the fact that unlike male sexual appetite, which is often encouraged and admired by societal norms, its female counterpart has been discouraged and suppressed— sometimes violently—making it difficult to get an honest assessment.

This double standard, although a cultural artifact, is likely based on human biology, specifically a trait that is shared by all mammals: internal fertilization. This anatomic reality has imposed a deeper and strictly biological double standard with respect to confidence in a parent's genetic relatedness to her versus his offspring. A woman is guaranteed to be the mother of a baby that emerges from her body, whereas absent DNA fingerprinting, a man cannot be sure. In short: mommy's babies, daddy's maybes. The consequence of this asymmetric uncertainty has been a near-universal cross-cultural pattern whereby men are especially concerned to restrict the sexual behavior of women, to a degree greatly exceeding women's efforts to restrict the sexual behavior of men. After all, male philandering does not in itself threaten a woman's genetic legacy, whereas female infidelity raises the possibility that her male partner may unknowingly end up rearing another man's child.

There is, nevertheless, substantial evidence all pointing toward polyandry as a female propensity, albeit not quite as flagrant as the male preference for polygyny—if only because male sexual jealousy has forced women to be especially secretive about their sex lives. Although overt polyandry—the mating system in which one female establishes an obvious mating relationship with multiple males—is rare, covert polyandry, in which females have more than one lover, seems to be the norm.

Studies of animal behavior, of the sort conducted by this author, encountered a similar disparity when it comes to the outright manifestations of animal nonmonogamy. Researchers have long been aware that male sexual gallivanting is common in the animal world, but we only came to appreciate the extent of its female counterpart when DNA fingerprinting became available for animal studies in the mid-1990s. To the surprise of many of my—mostly male!—colleagues, we found that even many bird species, long thought to be reliably monogamous, aren't. Depending on the species, anywhere from 10 to 70 percent of nestlings may be fathered by a male other than the mother's identified social partner. Social monogamy and sexual monogamy are not the same.

It immediately became apparent that the reason so many attentive field biologists had been fooled for so long is not that our female subjects were colluding to keep researchers in the dark, but rather because they have been selected to deceive their "husbands" as to their fidelity. And the primary reason for this appears to be that when male birds, for example, are experimentally induced to discover their mate in flagrante, these males are liable to cease provisioning or protecting the offspring—the avian equivalent of refusing to pay child support. There is no reason to think that female human beings are any less subtle or adroit.

Here, in brief, is some of the suitably subtle evidence for women's hidden polyandry.

Women conceal their ovulation. Unlike, say, chimpanzees, in which ovulation is clearly signaled by a bright pink anogenital swelling, the precise time of a woman's ovulation is a secret closely guarded, typically even from the woman in question. There are many hypotheses for this unusual situation, one being that by virtue of concealing the precise days when fertilization is most likely, ancestral women freed themselves from especially strict sexual oversight during those days, thereby gaining the opportunity to mate with other men, unbeknownst to their identified male protector/oppressor.

Similarly, women are also unusual among mammals in lacking a clear-cut estrous ("heat") cycle. The result is that far from being somehow at the mercy of their hormones with respect to their sex lives—an accusation that is more accurate with respect to men—women can exercise considerable choice, such that their private sex lives can, when and if desired, be divorced from their public social commitments.

Despite this freedom, and not inconsistent with it, evolutionary psychologists have found that women, during their ovulatory phase, are especially attracted to men who have a distinct pattern of body build and overall health consistent with high testosterone and, thus, what evolutionists identify as "good genes."[5] This frees women to follow a two-part reproductive strategy: establish a social bond with a man likely to be a responsible and reliable partner in resource acquisition and childrearing, but when possible and if so inclined, mate on the sly with men they find sexually attractive, where "attractiveness" ultimately translates into "possessing traits suggesting that their genetic offspring will be particularly successful."

The adaptive significance (evolutionary payoff) of female orgasm has long been a mystery, given that orgasmic women are no less successful in terms of their reproductive fitness than are their nonorgasmic contemporaries. However, when considered in the context of polyandry, two hypotheses for female orgasm make sense. For one thing, orgasm could serve as a way in which a woman's body speaks to her mind, confirming whether a sexual partner is a potentially good long-term match. And for another, orgasm could be rewarding not only for the woman involved, but also for her partner, thereby providing confidence as to her likely fidelity in the future—while, by allaying male suspicion, giving her the opportunity to be exactly the opposite.

Finally, male sexual jealousy is essentially a cross-cultural human universal. Such a widespread, deep-seated, and frequently even violent response wouldn't have been selected for if there weren't reasons for it, namely, the bio-logical male fear of being cuckolded and thus induced to rear another man's offspring (something that natural selection would strongly act against). Freud pointed out that we don't generally have social inhibitions against doing something that we wouldn't be inclined to do. There are, for example, no cultural prohibitions against eating your own feces. By the same token, men from Singapore to Saskatchewan wouldn't engage in strenuous efforts to prevent their mates from mating with other men if those mates were not, at least on occasion, inclined to do so.

Considering all of the above material, there is no question that a hypothetical visitor from one of those Trappist-1 exoplanets, examining the human species for the first time, would conclude that *Homo sapiens* is definitely not inherently monogamous. Here, then is another paradigm that deserves to be discarded.

This does not mean, however, that monogamy itself is ready for the ash-heap of cultural history. Rather, what merits overthrow is the simplistic assumption that monogamy, because natural, is necessarily easy, once we have met the right person. This myth of monogamy has caused immense distress, as well-meaning, well-behaved individuals have been blindsided by their own nonmonogamous inclinations, which often arise after they have committed themselves to someone else. Someone marinated in the myth of monogamy is liable at such a point to doubt his or her existing relationship, interpreting their perfectly natural temptation as a statement that they must not be suitably mated in the first place . . . or else they wouldn't be tempted.

Alternatively, they might perceive themselves as somehow morally flawed and a "bad" person, whereas the fact that they occasionally experience such inclinations means that they should be congratulated for being a genuinely healthy mammal. Whether they act on those inclinations is, of course, another matter, and one best left to each person's conscience and ideally, collaborative understanding with his or her partner. As the activist and advice columnist Dan Savage suggests, many people fall off the monogamy wagon but then climb back on, ending up being "monogam-ish." It's normal.

Among the avoidable tragedies of buying into the monogamy myth is when a victim reacts to his or her spark of interest in someone other than their designated partner (or to indications of similar interest on the part of said partner), by giving up on the whole enterprise, proclaiming—usually with despair—that they're "just not cut out for monogamy." The reality is that *no one is cut out for monogamy*. It is very rare, not only among human beings, but among other animals, too.

In the movie *Heartburn*, Meryl Streep's character complains to her father about her husband's infidelity. "You want monogamy?" he responds. "Marry a swan!" It is now clear, however, that even swans aren't reliably monogamous. My favorite and most reliably monogamous animal species is *Diplozöon paradoxicum*, a parasitic flatworm that inhabits the gills of freshwater fish. Among these creatures, male and female meet while adolescents, whereupon their bodies literally fuse together, and they remain sexually faithful until death. (Even then, it appears, they do not part.) For the rest of us, monogamy remains a very real and often a highly desirable possibility, but one that— precisely because it is unnatural—requires hard work, but also offers genuine rewards.

WHAT ARE THOSE rewards?

The main one appears to be biparental child care. Looking across the mammalian world, there is a consistent correlation between monogamy and male cooperation in rearing offspring. As already described, getting females to invest in their offspring is generally not a heavy evolutionary lift, since their genetic connection to their own children—hence, their inclination to invest maternally—is strong. For males, things are different. Indeed, so far as we know, this is the major (perhaps the only) reason why male mammals don't lactate. Given that their mates have literally carried the reproductive burden up until parturition, it would seem only fair for a mother's partner to give her a break at this time. But this assumes that males can do more to enhance their fitness by investing in infants to whom they might not be related than by seeking to enhance their situation in other ways, such as by maintaining their territory, competing with other males, wooing other females, and so forth. At the same time, paternal care can be immensely valuable for certain species, especially those in which neonates are helpless at birth and require long-term care and attention. This is precisely the situation for human beings, a species among which single-parenting is obviously possible, but unquestionably difficult.

Enter, that human frenemy, monogamy. It is the best, perhaps the only reliable way to tip the balance in favor of paternal involvement in care of offspring. This might well be the reason that monogamy has become so highly prized in much of the modern world, despite the fact that it goes against some of our more natural inclinations. Another possible reason for monogamy's persistence derives from the counterintuitive observation that men actually benefit more from monogamy than do women. Here's how it works.

Men often assume that they would have been happy in the "good old days" of prehistoric polygyny, each one happily ensconced as a successful harem-master. But most of them are wrong, analogous to those frauds who claim to have been Julius Caesar or King Tut in a previous life—whereas statistically, they are much more likely to have been among the nameless centurions who perished in Gaul, or one of the numerous, anonymous, woebegone slaves who labored and died constructing a pyramid. In any polygynous system (human or otherwise), the success of a limited number of harem-keepers occurs at the expense of a large number of failed wannabes, the number of whom is simply a function of the size of the average harem. If, for example, an average harem consists of ten females mated to one male, then each successful male mandates that on average there are nine who fail.

Polygyny is therefore a prescription for social chaos, since these abundant, socially excluded, and sexually frustrated males are constant bringers of bedlam, regularly attempting to overthrow and replace the dominant male. This leads to the suggestion that monogamy may have arisen as a kind of societal compromise whereby dominant men agreed to abide—at least in theory—by monogamous rules, foregoing their access to multiple females and guaranteeing a wife for most of these otherwise excluded males, in return for a degree of social peace and quiet.

Scholars are not all agreed that this "trade-off" hypothesis explains the widespread adoption of monogamy, even if we limit its purview to the Western world. But since we are speculating, here are some additional hypotheses, not for the ubiquity of monogamy but rather for some of the consequences of the fact that we are not natural born monogamists. Interpersonal violence is not only a cross-cultural universal but also something that is especially characteristic of men; moreover, it is particularly true of young men, and is overwhelmingly directed toward other young men. This suggests that our polygynous heritage is substantially responsible. The maleness of violence is so consistent that it is rarely examined, and yet, here is a stunning fact: if we could somehow do away with male-generated violence, we would pretty much eliminate violence altogether.

Here is another hypothesis, first elaborated in detail by Hector Garcia, in his provocative and convincing book, *Alpha God*.[6] The idea, in brief, is that our species was predisposed to monotheism by our ancient experience of polygyny. Although only a hypothesis, it makes sense of many aspects of monotheism. For example, the "one god" of the Abrahamic religions is nearly always imagined to be male. Also very big, very powerful, and inclined to violence, such that disobedience is not only dangerous, but liable to be lethal. He is, moreover, a "jealous" god—often depicted as concerned about "his" sexual exclusivity, as in "Thou shalt have no other gods before me," and analogizing errant Old Testament Israel to a woman who has transgressed by lusting after others. Like a dominant alpha male, god is typically seen, as well, as providing not only protection from enemies but also sustenance in times of need. If our species had been monogamous, and if (admittedly, a big "if," at least for believers) god has been fashioned after our image rather than vice versa, god would presumably have been imagined very differently.

In short, monotheism may very well have arisen in response to polygyny, not monogamy.

Another controversial speculation based on our nonmonogamous past deals with the fraught question of genius, specifically, why so many identified geniuses have been men. The operative word here should be "identified," because there is no reason to think that men are actually more prone to being geniuses than are women. What is clear, on the other hand, is that social pressures have been such that women of talent have historically been prevented from reaching their potential. Similarly, it is also possible that exceptionally talented men have been more likely to be recognized simply because they are men, whereas comparable genius among women—even women able to overcome societal bias and achieve educational opportunities and the freedom to express their capacities—has been downplayed and seen as less valuable.

There is also this possible connection to our polygynous past: even though men are no smarter and no more creative than women, perhaps there is something about maleness that predisposes men to call attention to whatever potentially notable qualities they might possess. Like a male silverback gorilla thumping his chest, or a male bird of paradise flaunting his feathers, maybe men are simply more prone to advertising themselves, which might in turn lead to enhanced efforts to draw attention to whatever they write, paint, compose, sculpt, and so forth, as part of a characteristically mammalian inclination to attract notice and thus, in theory at least, mates. Call it "Portrait of the Artist as a Show-Off," especially male artists.

Since we are speculating, here is a final hypothesis, yet another potential consequence of our nonmonogamous ancestry. Could it be that homosexuality derives at least in part from the fact that our prehistoric mating pattern was both overtly polygynous and covertly polyandrous? Here is how it might have happened: With a relatively small number of dominant alpha men monopolizing more than their share of women, there would necessarily be an excess of unmated men. Those men who were heterosexually inclined would have had a hard life, competing with each other for opportunities to overthrow the harem-master and then obliged to maintain their status. By contrast, men whose needs were met homosexually would have had other men available as potential partners; moreover, their sexual preference wouldn't have evoked violent antagonism from harem-keeping males, because unlike heterosexuals, they wouldn't have constituted a threat to the harem-keeper's suzerainty.

Female homosexuality could also have been predisposed in such circumstances, in part because cowives would have had more access to each other than to the dominant male, and also because although harem-keeping males are often lethally intolerant of their wives taking heterosexual lovers, lesbian relationships would likely have been viewed more benignly. A remaining problem, however, is that even if the above pattern actually occurred, for either gay men or lesbian women to have been selected for, there would have had to be some way in which homosexual-predisposing genes actually promoted themselves into the future.

Regardless of why monogamy became the dominant domestic ideology, and despite the fact that it isn't "natural," and notwithstanding whether the above speculation for the roots of violence and of monotheism, homosexuality, as well as male displays of genius turn out to be valid,[7] there is simply no doubt that among the paradigms of human nature worth discarding, the naturalness of monogamy is high on the list. On balance, moreover, it benefits men, women, and children, and contributes to a more peaceful society, but at a price: namely, requiring both men and women to inhibit some of their inclinations. In *Civilization and Its Discontents*—his best book—Freud noted that civilization requires the suppression of our instincts, notably the "id," which effectively requires that people sign on to monogamy along with other restrictions on their freedom of action. According to Freud, the result is neurosis, although the alternative—fatherless children and violent chaos—is almost certainly worse.

So long as we are discarding myths, along with that of monogamy we are well advised to jettison the deeper illusion that what's natural is necessarily good, along with its converse, that whatever is unnatural must be bad. Natural foods, natural environments: there is much to be praised when it comes to naturalness. But there is much to be deplored, too: earthquakes, tsunamis, hurricanes, and tornados are generally natural and yet, often devastating and by most conceptions of the word, "bad." Similarly for most diseases. What is more natural than the bacteria that cause typhoid, or the virus that produces polio? More than two centuries ago, when Voltaire was appalled by the immense suffering caused by the Lisbon earthquake of 1755, he wrote his novel *Candide*, which satirized the naive "Panglossian" view that "all was for the best in this best of all possible worlds." The philosopher David Hume similarly emphasized that "is" does not necessarily imply "ought," that we aren't justified extrapolating from the way the world is to how it ought to be.

A similar argument is warranted when it comes to evaluating our own behavioral predispositions. Just because we may feel inclined—naturally—to do something, it does not follow that we ought to do it. Human beings are unusual in the animal world when it comes to our ability to do things that do *not* feel good, and/or to refrain from certain things that do. There are aspects of monogamy that indeed feel good; others, less so. Although monogamy isn't natural, it may nonetheless be good. Similarly, because it isn't natural, it isn't easy; it takes work. But some of the most admirable things that *Homo sapiens* accomplish are equally unnatural, and are achieved only by hard work: learning a second language or how to play the violin, getting toilet trained, not punching a jerk in the nose, not giving up when the going gets rough.

Some things are so unnatural as to be literally impossible: we cannot flap our arms and fly like a bird, or tell our kidneys to stop filtering our blood. But other things, despite being unnatural, like monogamy, are without doubt possible, as evidenced by the fact that many people are in fact monogamous. Having outgrown the misleading paradigm that monogamy is a "natural" human state, anyone wanting to be monogamous should if anything find it easier insofar as he or she understands that it will be difficult but may nonetheless be worthwhile.

As with various other consequences of seeing our species, and thus ourselves, as we really are, self-knowledge enhances the prospect of mindful self-control and thus, living our lives as we choose, whether in accord with our biology or in defiance of it.

Old paradigm: People are naturally monogamous, if only they find their ideal life partner.

New paradigm: Men are naturally polygynous, interested in multiple female partners, and women are naturally polyandrous. But both sexes are essentially free to be whatever they choose, particularly if they free themselves from the straitjacket that is the myth of monogamy.

NOTES

1. D. P. Barash, *Out of Eden: Surprising Consequences of Polygamy* (New York: Oxford University Press, 2016).
2. M. Konner, *Women After All* (New York: Norton, 2016).
3. I. Dupanloup et al. "A Recent Shift from Polygyny to Monogamy in Humans Is Suggested by the Analysis of Worldwide Y-Chromosome Diversity," *Journal of Molecular Evolution* 57, no. 1 (2003): 85–97.
4. Described in all its suitably salacious detail in Barash, *Out of Eden.*
5. R. Thornhill and S. Gangestad, *The Evolutionary Biology of Human Female Sexuality* (New York: Oxford University Press, 2008).
6. H. Garcia, *Alpha God* (Amherst, NY: Prometheus, 2015).
7. For substantially more supportive arguments, see Barash, *Out of Eden.*

13

War and Peace

NOT ALL OF our pre-evolutionary illusions place us in an unrealistically favorable light. One in particular—that human beings are instinctively predisposed to war—is a distinctly *unfavorable* idea, as well as distinctly untrue. Also, interestingly, it is one that has been promoted by many evolutionary biologists who should know better. The notion is that *Homo sapiens* owes his and her nature (especially "his") to an ancient war-making proclivity that is an irrevocable mark of Cain on our species.[1]

This chapter takes issue with this perspective, but also with the converse illusion, whereby some specialists (social scientists in particular) have engaged in an equally misleading (albeit perhaps less dangerous) "pacification of our past," portraying human beings as angelically nonviolent.

First, our supposed war-loving instincts. There can be something oddly appealing about taking a dim view of human nature, something that has become unquestioned dogma among many evolutionary biologists. It is a tendency that began some time ago. When the Australian-born anthropologist Raymond Dart discovered the first australopithecine fossil in 1924, he went on to describe these early hominids as "Confirmed killers: carnivorous creatures that seized living quarries by violence, battered them to death, tore apart their broken bodies, dismembered them limb from limb, slaking their ravenous thirst with the hot blood of the victims and greedily devouring living writhing flesh."[2]

This lurid perspective has deep antecedents, notably in certain branches of Christian doctrine. According to the zealous sixteenth-century French theologian John Calvin, "The mind of man has been so completely estranged from God's righteousness that it conceives, desires, and undertakes, only that which is impious, perverted, foul, impure and infamous. The human heart is so steeped in the poison of sin, that it can breathe out nothing but a loathsome stench."[3] Misanthropy can also be purely secular, as in this observation from Aldous Huxley:

> The leech's kiss, the squid's embrace,
> The prurient ape's defiling touch:

And do you like the human race?

No, no much[4]

It is bad enough for the religious believer to be convinced of humanity's irrevo-cable sinfulness, punishable in the afterlife. And readers of the present book have already encountered nonreligious paradigms concerning human nature that are less than complimentary. But it is one thing to take an objective, science-based and clear-eyed look at *Homo sapiens*, warts and all, and another for those who speak for sci-ence and reason to promote a theory of human nature that threatens to become a hurtful self-fulfilling prophecy. For example, in his influential book, *African Genesis* (1961), Robert Ardrey described humans as "Cain's children." Ardrey wrote that *Homo sapiens* "is a predator whose natural instinct is to kill with a weapon. It is war and the instinct for territory that has led to the great accomplishments of Western Man. Dreams may have inspired our love of freedom, but only war and weapons have made it ours."[5]

No other animal species has been observed using weapons—that is, artificial exten-sions of their bodies—to kill or injure conspecifics. There have been, interestingly, accounts of chimpanzees using a tool to kill other vertebrates: sharp sticks were poked into tree holes, skewering small sleeping nocturnal primates known as galagos.[6] But here, the "weapons" were wielded primarily by adolescent females, which is not con-sistent with expectation if such behavior were an antecedent to warfare.

Nonetheless, the drumbeat that argues for war as a defining feature of the human condition has, if anything, increased in recent decades, spreading beyond the evolu-tionary and anthropological worlds. An article in *The National Interest* titled "What Our Primate Relatives Say about War" answered the question "Why war?" with "Because we are human."[7] At about the same time, a piece in *New Scientist* asserted that warfare has "played an integral part in our evolution"[8] and a research report in the journal *Science* claimed, "death in warfare is so common in hunter-gatherer societies that it was an important evolutionary pressure on early *Homo sapiens*."[9] And here is how New School philosopher Simon Critchley began his review of John Gray's *The Silence of Animals* (2013) in the *Los Angeles Review of Books*: "Human beings do not just make killer apps. We are killer apes. We are nasty, aggressive, violent, rapacious hominids."[10]

Then there is the American anthropologist Napoleon Chagnon, who devoted decades to studying the Yanomamo people of the Venezuelan/Brazilian Amazon. His best-selling book *The Fierce People* (1968) has been especially influential in enshrin-ing an image of tribal humanity as living in a state of "chronic warfare." Chagnon has been the subject of intense criticism, although, to my mind, there is simply no question about the empirical validity and theoretical value of his research. He was nonetheless grossly abused by much of the academic anthropology establishment, accused of mis-deeds of which he was entirely innocent, animus that derived in large part because his findings—that the Yanomamo are in fact belligerent and warlike—went against the ideological preferences of many of his colleagues. (Herein lies an important tale, fit perhaps for another account: the extent to which personal politics often enters into the question of whether human beings are inherently violent.) In any event, after consid-erable controversy, Chagnon's research has been widely accepted, perhaps too widely. Or rather, it has been overgeneralized.

In a field (call it evolutionary psychology or, as I prefer, human sociobiology) that has often been criticized for a relative absence of hard data, Chagnon's findings, how-ever politically distasteful, have been welcome indeed. Among these, one of the most convincing has been Chagnon's demonstration not only that among the Yanomamo intervillage "warfare" is frequent and lethal but also that Yanomamo men who have killed other men experience significantly higher reproductive success—evolutionary fitness—than do nonkillers.[11] His data appear reliable and robust, and yet, a reassess-ment of those findings revealed that those who have killed—labeled *unokai* in the Yanomamo tongue—also tend to be, on average, ten years older than those who have not.[12] It stands to reason that older men would have more descendants, regardless of their violent reputations. Moreover, a reliable assessment of the fitness consequences of murderousness would seem to require looking not only at those who killed and lived to leave descendants but also those who met a violent and untimely end; otherwise, the result is like trying to evaluate the lethality of the recent Ebola epidemic by reporting on those who survived it.

I have a growing sense of discomfort about the way that Chagnon's Yanomamo research has been interpreted and the inferences that have been drawn from it. I fear that many of my colleagues have failed, as previously have I, to distinguish between the relatively straightforward evolutionary roots of human violence and the more com-plex, multifaceted, and politically fraught question of human war. Moreover, research on another violent society, the Waorani of South America (who have a higher homi-cide rate, in fact, than the Yanomamo), after examining the life histories of 85 identified warriors and 121 elders, concluded that "more aggressive men, no matter how defined, do not acquire more wives than do milder men, nor do they have more children, nor do their wives and children survive longer."[13] To be blunt, violence is almost certainly deeply entrenched in human nature; warfare, not so much. Probably, in fact, not at all.

A fascination with the apparent correlation between Yanomamo violence and male fitness (i.e., evolutionary success, not simple physical prowess) has blinded us to the full range of human nonviolence, causing us to ignore and undervalue realms of peace-making in favor of a focus on exciting, attention-grabbing, and ultimately misleading patterns of war making. As an evolutionary scientist, I have been enthusiastic about identifying the adaptive significance—the evolutionary imprint—of apparently universal human traits. For a long time, it seemed that Chagnon's finding of the reproductive success of Yanomamo men who were killers was one of the most robust pieces of evidence for this. Now I am not so sure, and this chapter is—among other things—my *mea culpa*.

THERE HAS ALSO been a tendency among evolutionary thinkers to fix on certain human groups (especially nontechnological and violent ones) as uniquely revelatory, not simply because the research about them is robust and because they are thought to offer special insights into our evolutionary antecedents but also because their stories are both riveting and consistent with our preexisting expectations. In addition, they are just plain fun to talk about, especially for men. Remember, too, the journalist's edict, "If it bleeds, it leads." You are unlikely to see a newspaper headline announcing, "France and Germany Did Not Go to War Today," whereas a single lethal episode, anywhere in the world, is readily pounced on as news.

Language conventions speak volumes, too. It is said that the Bedouin have nearly one hundred different words for camels, distinguishing between those that are calm, energetic, aggressive, smooth-gaited, rough, and so forth. Although we carefully identify a multitude of wars—the Hundred Years' War, the Thirty Years' War, the American Civil War, the Vietnam War, and so forth—we don't have a plural form for peace. Wars, yes; "peaces," no. It makes evolutionary sense that human beings pay special attention to episodes of violence, whether interpersonal or international: they are matters of life and death, after all.

But when serious scientists do the same and, what is more, when they base "normative" conclusions about the human species on what is simply a consequence of their selective attention, we all have a problem. The most serious problem with the belief that war is part of human nature is one familiar to many branches of science: generalizing from one data set—however intensively studied—to a wider universe of phenomena. Academic psychologists, for example, are still reeling from a 2010 research report from the University of British Columbia that found the majority of published psychological research derives from college students who are "Western, Educated, Industrialized, Rich, and Democratic"—in short, WEIRD.[14] Although the Yanomamo are not weird—they are people, like ourselves—they are only one of a large number of traditional human societies who differ from each other in many ways. Given the immense diversity of human cultural traditions, any single group of *Homo sapiens* (including but not limited to the college sophomores so enamored of social psychology researchers) must be considered profoundly unrepresentative of the species as a whole.

Just as the Yanomamo can legitimately be cited as violence-prone—at both the individual and group level—many other comparable traditional peoples do not engage in anything remotely resembling warfare. These include the Batek of Malaysia, the Semang of Malaysia, the Mardu of Australia, a half-dozen or more indigenous South Indian forager societies, and numerous others, all of whom are no less human than those regularly trotted out to "prove" our inherent war-proneness.

In the Dark Ages of biology, taxonomists used to identify a "type species" thought to represent each genus, but the idea no longer has any currency in biology. The great evolutionary biologist Ernst Mayr effectively demonstrated that statistical and population thinking trumps the idea of Platonic "types," independent of the actual diversity of living things, not least *Homo sapiens*. Yet many anthropologists (and biologists, who should know better) seem to have fallen into the trap of seizing on a few human societies and generalizing them as representative of humanity as a whole. Regrettably, this tendency to identify "type societies" has been especially acute when it comes to establishing the supposed prevalence of human warfare.

It is remarkable that otherwise accomplished scholars should make such a beginner's error, equivalent to seeing a robin and concluding that all birds have red breasts. The Yanomamo are indeed quite violent, but lots of other human groups are not. And interpersonal violence is not the same as warfare, which involves organized group lethality.

Working with the psychologist Patrik Söderberg, the anthropologist Douglas Fry carefully reviewed evidence for violence among current mobile foraging societies. Out of 21 such societies for which high-quality data were available, Fry and Söderberg identified 148 cases of lethal violence. Of these, fully 69 came from one group, the Tiwi people of Bathurst and Melville Islands, off the Australian coast. Altogether, 55 percent of the lethal instances consisted of one killer and one victim; hence, they exemplified homicide rather than war. In 23 percent, more than one person participated in killing one person, many of which involved a coordinated effort at eliminating a single violent individual; hence, they were closer to a kind of nonjudicial but intrasocietal "police action." Only 22 percent involved multiple perpetrators and multiple victims. Ten of the twenty-one societies evinced no lethality generated by more than one person, and in three, there were no recorded cases of lethality.[15]

In his justly admired book *The Better Angels of Our Nature* (2011), the evolutionary psychologist Steven Pinker made a powerful case that human violence—interpersonal as well as warring—has diminished substantially in recent times. But in his eagerness to emphasize the ameliorating effects of historically recent social norms, Pinker exaggerated our preexisting "natural" level of war-proneness, claiming that "chronic raiding and feuding characterized life in a state of nature."[16] Reality is otherwise. As recent studies by the anthropologist Douglas Fry and others have shown, the overwhelmingly predominant way of life for most of our evolutionary history—in fact, pretty much the only one prior to the Neolithic revolution—was that of nomadic hunter-gatherers. And although such people engage in their share of *interpersonal* violence, warfare in the sense of lethal, organized intergroup violence, is almost unknown.

After examining fifty well-studied forager societies, the primatologist Christopher Boehm concluded that when individuals persistently violate group norms—as by excessive bragging, bullying, cheating, lying, stealing, or murdering—a range of social sanctions are nearly always employed, including ostracism, public shaming, and so forth. Rarely, and on occasion in the event of an acknowledged murderer, such an individual might be killed. But nonviolent alternative punishments are consistently preferred and are much more common.[17]

Moreover, evidence from paleoanthropology shows that such violence directed by one group against other groups was pretty much nonexistent in the past as well, having emerged only with early agricultural surpluses and the elaboration of larger-scale, tribal organization, complete with a warrior ethos and protomilitary leadership. It thus appears that war is a comparatively recent cultural add-on to the human repertoire, acquired within the last 10,000 years or so, as a result of several factors, including the invention of agriculture—with subsequent accumulation of valuable material resources, which lend themselves to being stolen as well as defended—along with the attainment of elaborate hierarchical social structures, plus increasingly effective technologies for communication, coordination, and killing.

Nonetheless, in *The Social Conquest of Earth* (2012), the biologist Edward O. Wilson calls warfare "humanity's hereditary curse."[18] I applaud both Pinker's and Wilson's distaste for war, but I wish they had thought more deeply and consulted the cross-cultural and archaeological evidence more carefully before jumping on the "war has always been with us" bandwagon. Human history demonstrates that warfare is recent, not ancient, and certainly not fundamental to most of our evolutionary story. But what about our closest ape relatives?

Savannah baboons from East Africa were among the first free-living nonhuman primates to be studied, and—just by chance—they are also among the most aggressive and violent. This may well have predisposed subsequent primate research to emphasize aggressiveness and violence.

As with human societies, we have no living ancestors among currently existing nonhuman primates, no matter how "primitive" they may appear. None of today's great apes are our direct ancestors, although we share a common antecedent with today's chimpanzees, bonobos, gorillas, orangutans, and gibbons. The social behaviors of the latter three are all dramatically different from those of modern human beings, while those of chimpanzees and bonobos are almost exactly opposed to one another: chimpanzees are known to engage in violent, group-level encounters, complete with search-and-destroy missions that conjure images of human skirmishing and outright warfare. Bonobos, on the other hand—genetically no more distant from *Homo sapiens*—do nothing of the sort, and are renowned for making love, not war.

Choosing either species as a model for "natural" human beings—as with choosing a particular traditional human society as especially revelatory of our past—says more about the ideology of the person doing the choosing than about the underlying biology of human beings generally. (It is worth noting that just as war-prone chimpanzees are regularly referenced by the "war-is-natural" crowd, bonobos appear as a predictable trope employed by those commentators who prefer "make-love-not-war" as the human default predisposition.)

Chimpanzees occasionally hunt and kill other primates; bonobos do not. The role of predation in forming the human psyche is an open question, with some anthropologists emphasizing the formative influence of "man the hunter," and others arguing that "woman the gatherer" prehistorically provided many more calories—and, thus, generated more influence on our early evolution. Either way, it is important to distinguish between interspecies predation and intraspecies aggression; although the former is certainly violent, predatory species such as wolves and lions employ entirely different behavior patterns when obtaining their prey as compared to when engaging in violent competition with other wolves or lions.

WHAT, THEN, IS the biologically given state of *Homo sapiens* when it comes to violence and war? Unfortunately for those who like their answers simple, reality is ambiguous . . . or rather, ambidextrous, in that it points in two conflicting directions. Our species is certainly *capable* of violence at both the individual level (e.g., assault, rape, homicide) and that of groups (war). But a capacity is a far cry from a necessity, which implies a predisposition simmering just below the surface, urgently seeking opportunities to burst out. There is no evidence that human beings who have lived a consistently nonviolent life feel an eventual need to commit mayhem at the behest of their frustrated genes and, moreover, to do so as organized groups in conflict with other such groups. At the same time, there is abundant evidence that at the level of whole societies, people are quite capable of eschewing war—because numerous societies have done just this.

Nonetheless, it is equally evident that natural selection has equipped our species with a predilection for violence under certain circumstances—namely, when resource competition is high (e.g., for food, mates, territory) or under a variety of social conditions (e.g., when status issues are sufficiently intense, as a result of hierarchical manipulation or instability), and that such vulnerability is especially great among an identifiable subset of humanity (young adult males). It also deserves emphasis that whereas interpersonal violence is directly associated with relatively simple neurobiological causation—involving such readily identifiable brain regions as the limbic system and particular neural nuclei and transmitter hormones—war is quite different, requiring elaborate cognitive processes, extensive planning and—ironically, perhaps—substantial social cooperation, at least among those on the same side.

The framers of the US government system were acutely aware of the difference between individual violence and group war, and therefore of the risk arising from the fact that political leaders often enjoy personal benefit when they lead their fellow countrymen to war. In the fourth *Federalist* Paper, John Jay explained why for this reason it should be made politically difficult for the President to do so. "It is too true," he wrote,

> however disgraceful it may be to human nature, that nations in general will make war whenever they have a prospect of getting anything by it; nay, absolute monarchs will often make war when their nations are to get nothing by it, *but for the purposes and objects merely personal,* such as thirst for military glory, revenge for personal affronts, ambition, or private compacts to aggrandize or support their particular families or partisans. These and a variety of other motives, which affect only the mind of the sovereign, often lead him to engage in wars not sanctified by justice or the voice and interests of his people.

Just as a "wall of separation" was constitutionally constructed between church and state in the early United States, the requiring the House of Representatives to approve a declaration of war by a two-thirds majority reflected explicit recognition by this country's founders that great care must be taken when bridging the gap between individual predilection for violence (especially when it's of the "Let's you and him fight" sort, and a group undertaking.

Also at issue here is the matter of "levels of selection," the extent to which natural selection operates among genes as opposed to individuals, or even among groups. Although the great majority of biologists—including myself—agree that natural selection is more potent in proportion as its level is fine-grained (e.g., more effective among individuals than among groups, and more so among genes than among individuals), in some quarters there has been an increase in the perceived legitimacy of so-called group selection, especially in the case of human beings.[19]

On the one hand, arguments against group selection remain cogent, especially for cases in which individual, selfish benefit is likely to swamp any progroup, "altruistic" inclinations. This is because for them to persist, group-level adaptations would have to cause altruistic groups to outreproduce nonaltruistic groups by a difference that is greater than the extent to which selfish individuals within groups have an advantage over altruists within those same groups. The problem, in short, is that when it comes to reproducing, individuals experience far greater variability than do the groups to which they belong. Moreover, mathematical models suggest that gene flow between groups would further facilitate the spread of any selfishness-predisposing alleles by introducing them into groups composed of other-oriented altruists who are—because of their altruism—liable to be particularly vulnerable to such invasion. Current data for extant nomadic foragers indicates that in fact this pattern may well have been predominant among ancestral human populations as well.

On the other hand, it is at least possible that despite its acknowledged rarity in the natural world, selection at the level of groups might have been somewhat influential in human evolution, if only because human beings seem likely to have been especially effective in policing "defectors" from prosocial group norms, thereby possibly ameliorating the tendency for selection to favor selfish individualists over group-oriented cooperators. The scientific jury is currently out with respect to this question, especially in the case of *Homo sapiens*.[20]

Whether a clear verdict will eventually be reached or a hung jury results, the issue is very relevant for our purposes. If group selection is eventually exonerated, it would provide yet another potential avenue whereby prosocial, altruistic cooperation could have evolved. (As we'll soon see, however, group selection itself—like altruism and selfishness—is a two-edged sword when it comes to the question of human warlike versus peaceful tendencies.) In theory, altruism could evolve at the level of groups if benevolent altruists conveyed sufficient benefit to the group as a whole that such a group, including its constituent altruist genes, prospered at the expense of other groups that lacked comparable altruists. If so, then prosocial, altruistic traits could evolve even if they were not preferentially directed toward genetic relatives—that is, even if they were not the outcome of "selfish genes." (More about this in the next chapter.)

There are, however, a couple of significant hurdles that must be overcome in order for group selection to emerge as a significant evolutionary force. As already noted, the fitness benefit conveyed to entire groups must be sufficient to compensate for the fitness loss necessarily borne by individual altruists within their groups. For most living things, this appears to be an insuperable obstacle. In addition, unless altruistic beneficence is preferentially directed toward genetic relatives (in which case it is indistinguishable from kin selection) group selection for altruism seems doomed to fall victim to "free riders" within groups who take advantage of the altruism of others without having to pay the cost of being altruistic themselves.

It must be noted, however, that even though the pendulum of scientific opinion continues to swing strongly against group selection as a robust driver of evolution among most species, it might have operated among *Homo sapiens*—perhaps uniquely among organisms. This is because human beings may also be unique among living things in their ability to establish and enforce group norms, thereby protecting against excessive payoffs to free-riders. It is accordingly possible that some of the traits that uniquely characterize human beings (notably, religion), that impose a cost on participants, and that are otherwise difficult to explain in evolutionary terms might have evolved in response to a selective push operating at the level of groups. It has plausibly been suggested, for example, that the adaptive payoff of religion may derive from advantages in coordination and social solidarity experienced by groups as a result of the religious affiliation of their individual members.

Lest peace advocates celebrate prematurely, however, let's note, first, that kin selection and reciprocity may well prove sufficient to explain human adaptations otherwise attributed to group selection. Moreover, even if group selection is eventually found to be fundamental to human evolution, it is yet another example of that two-edged sword mentioned earlier. Thus, among those group-level benefits that could have accrued to our ancestors, it is entirely possible (perhaps even likely) that success in competition with other groups would have loomed large. If so, then intergroup hostility—analogous to warfare—might have been among the significant threats experienced by our early prehuman and human ancestors. Groups sharing the same religion, within which individuals are more likely to sacrifice for the benefit of others, and that displayed greater social cohesion as a consequence, might have profited via these characteristics when it came to violent competition with other groups. Insofar as group selection was instrumental in sculpting the human species, the result was therefore as likely to predispose *Homo sapiens* toward competitive war as toward cooperative peace.

At the same time, a strong case can and has been made that nomadic forager social systems in particular predispose against violent intergroup competition, for several reasons. For one, population structure of extant groups suggests that individuals often have close genetic relatives in neighboring groups, which would mitigate against violent conflict. For another, when they do arise, conflicts typically are interpersonal—between two men, for example, over a woman—rather than among groups. In addition, it is common for competition over variable and limited resources to result in reciprocal sharing and cooperation rather than prototypical warfare. The primary threat to ancestral human beings may also have been posed by nonhuman animal predators rather than other human beings. The greatest likelihood is that proto-, pre-, and early humans faced numerous challenges: from other organisms, other human beings, as well as numerous and complex environment-based difficulties, with no single one having been clearly determinative.

President Harry Truman is reputed to have lamented the absence of one-armed economists, as follows: when he asked his economic advisers for the likely outcome of a particular government policy, he would be given a possible scenario, after which he would be told, "But on the other hand . . . " Something similar results when attempting to evaluate the impact of human evolution acting on early group structure (whether via group or individual selection). Thus, studies of pretechnological "war" among horticulturalists such as the Yanomamo and Mundurucu suggest relatively high levels of interindividual as well as intergroup violence and mortality. Once again, the situation seems more complicated, bifurcated, and perhaps genuinely ambiguous than partisans on either side of the "primitive peace/primitive war" divide would prefer.

When it comes to human aggression, violence, and war, there simply is no unitary direction impelled by evolution or mandated by our biology. On the one hand, we are capable of despicable acts of horrific violence; on the other, we evince remarkable compassion and self-abnegation. Our selfish genes can generate a wide array of nasty, destructive, and unpleasant actions; and yet these same selfish genes can incline us toward altruistic acts of extraordinary selflessness. It is at least possible that our remarkably rapid brain evolution has been driven by the payoff derived by successful warlike competition with other primitive human and humanoid groups. But it is equally possible that it was driven by the payoff associated with cooperation, coordination, and mutual caretaking or both.

It is relatively easy to develop mathematical models whereby animals and people who are adroit at these tasks—along with genes that predispose in such directions—will be favored by natural selection over alternative individuals and alleles that are comparatively more bellicose. Moreover, even as organized warfare is new to the human experience and therefore liable to be culturally induced rather than biologically based, behavioral systems of restraint are old, shared by numerous animal species, and therefore likely to be if anything more deep-seated in our nature. Even in a war-riven world, actual wars are much rarer than are examples of nonviolent conflict resolution; the latter happens every day, among nations no less than between individuals.

We could never be as altruistic as worker honeybees, or as solitary as deep-sea angler fish. There are no human societies whose members are expected to have sex a dozen times each day, or not at all. Or in which people are expected to consume 25,000 calories per day, or to stop eating—or breathing—altogether. Clearly there are biological restraints on our behavioral repertoire, and culture reflects these restraints. But there is a limit to our genetically influenced proclivities—and this, please note, is written by an evolutionary biologist who has been accused of hypothesizing genes for just about everything.

A useful distinction in this regard is between evolved adaptations and capacities. Language is almost certainly an adaptation, something that all normal human beings can do, although the details vary with circumstance. By contrast, reading and writing are capacities, derivative traits that are unlikely to have been directly selected for, but have developed through cultural processes. Similarly, walking and probably running are adaptations; doing cartwheels or handstands are capacities. Interpersonal violence is a human adaptation, not unlike sexual activity, parental care, communication/manipulation, and so forth. It is something we see in every human society. Meanwhile, war—being historically recent, as well as erratic in worldwide distribution and variable in detail—is almost certainly a capacity. And capacities are neither universal nor mandatory.

Let me be clear. Violence is widespread and, sadly, deeply human, just as the adaptation for violence under certain circumstances is similarly ingrained in many other species. But war is something else. It is a capacity, one that involves orchestrated group-based lethal violence. Thus, it deserves to be distinguished from rivalry, anger, "crimes of passion" or revenge, or other forms of homicide. To engage an absurdly positive simile: violence is like marriage, in the sense that some sort of process whereby adults solemnize their relationship appears to be a cross-cultural universal, and is a likely candidate for being an adaptive part of human nature. War, on the other hand, is like arranging a wedding with a bridal shower or bachelor party, plus adding a hotel ball-room, an orchestra, a four-course meal, and dancing. It is safe to assume that neither employing a photographer, serving a multitiered wedding cake, enlisting bridesmaids, nor tying baby shoes to the bumper of the newlyweds' car derive from the human genome, although people are capable of doing all these things. By the same token, plain old interpersonal violence is a real, albeit regrettable, part of human nature. War is even more regrettable, but is no more "natural" than a bridal shower or the assembly line used to construct a stealth bomber.

WE CAN TAKE some solace from the fact that war is considerably less inscribed on our genes than is individual violence, if the former is there at all (which I very much doubt). Nonetheless, the connection between personal homicide and societal war is closer than might be thought—and not simply because societal war involves the participation of individuals. It is rare indeed when a war breaks out that—as many admirable pacifists have hoped—no one shows up. Worse yet, in large-scale societies outfitted with compa-rably large-scale weapons and with large numbers of soldiers ready to act at the behest of perhaps one leader, the devastating prospect beckons that a given leader (or small deci-sion-making group) is motivated by the equivalent of personal homicidal inclinations—made even more likely in modern times when the war-makers don't typically have to go to war themselves—whereupon a vast group becomes mobilized for destruction.

The psychologist B. F. Skinner wrote, "no theory changes what it is a theory about; man remains what he has always been."[21] True enough for most phenomena: the solar system is as it was when Ptolemaic thinking held sway; nor did this change after Copernicus, Kepler, and Galileo came up with a more accurate descriptive theory. Gravity was gravity before and after Newton. But theories of human nature are dif-ferent. When I write or lecture about the social behavior and reproductive strategies of different marmot species, no sociopolitical implications are involved. But when considering violence, aggression, and/or war-making in humans, it makes a huge dif-ference whether we are thinking of the pacific Lepcha of the Himalayas or the fierce Yanomamo.

It's not that any particular theory changes human nature itself. The present book focuses, among other things, on the fact that human beings are not central to the cos-mos. If this is incorrect, and we really are the apple of the universe's eye, my assertions to the contrary will not demote us—except in our own self-esteem, and then, only temporarily. However, insofar as our expectations and thus our behavior changes, ideas about innate human war-proneness, for example, could have genuine consequences.

Thus, if we are convinced that Thomas Hobbes was correct and people are naturally inclined to be nasty and brutish (not only as individuals but also as organized groups), then it follows that violence and war are inevitable, and efforts toward peace are futile and we should all resign ourselves to a Hobbesian "warre of every man against every man." This has profound implications for our politics, including our sense of national budgetary priorities—how much to invest in, say, education and healthcare versus the police and the military—which in turn is liable to become a self-fulfilling prophecy. How many arms races and cycles of international distrust have been fed by a preexisting view that the other side is aggressive, potentially violent, and irremediably warlike, which in turn leads to policies and actions that further confirm such assumptions?

The danger, in short, of assuming that *Homo sapiens* has a "natural instinct" for war is that it can become a highly destructive self-fulfilling prophecy, not only closing off possible avenues of peaceful conflict resolution, but actually making war more likely. Nonetheless, a purportedly scientific view of anything—humanity's presumed instinct for warfare included—must stand or fall not on its social and political consequences but on its scientific credentials. And here, the "war is in our genes" perspective is not only ethically suspect but scientifically invalid.

Especially when they speak to matters of war and peace, theories about the fundamental nature of human beings are more insidious and influential than other evolutionary research, such as whether modern humans carry Neanderthal genes, or the potency of gene versus individual versus group selection, and so forth. It's not that we inhabit a postmodern world in which language and ideology construct reality. Rather, certain ideas have real effects on crucial topics such as national levels of military versus domestic investment and whether or not to go to war.

At risk of repeating: this is not to claim that scientific perspectives should be evaluated by their ideological, ethical, political, and social implications. Unlike matters of ideology, politics, ethics, or theology, science must be assessed only by the degree to which its fruits are, or are not, falsified; the confidence with which we can agree on their usefulness; and the extent to which they generate further testable ideas. But we must also remain alert when science risks being misused for ideological purposes, not to mention the subtle extent to which researchers can unintentionally bias their findings, simply by their choice of research subjects—as when one or a few human societies are taken as indicative of our entire species, or when certain findings are trumpeted as being more generalizable than others. This problem is magnified when the danger exceeds simply being misled on the "strictly scientific," to embracing consequences for social policy.

The human mind is drawn toward simple either/or statements—god versus the devil, cowboys versus Indians, you're either with us or you're with the terrorists—but reality is more nuanced and complex. H. L. Mencken once noted that to every question there is an answer that is simple, satisfying—and wrong. This applies particularly to the seemingly simple matter of whether human beings are "naturally" or "instinctively" aggressive or violent. Our human nature is neither Rousseauian nor Hobbesian; instead, both a devil and an angel perch on our shoulders, gesturing toward evolutionary predilections in both directions.

Even as human evolution has permitted and in some cases even encouraged the elaboration of violent and occasionally warlike behaviors, it has also promoted prosocial activities and inclinations, including but not limited to altruism, empathy, and numerous aspects of coordination—for example, for social learning, handicrafts and tool-making, constructing homes, animal enclosures, food storage systems, crop-raising and animal domestication, organizing daily and migratory movements, and so forth. An increasingly viable hypothesis suggests that the elaborate intellectual prerequisites for success in a highly social species exerted substantial selection pressure for the evolution of the uniquely high level of human intelligence itself. At the same time, such intelligence may also have set the stage for benevolent and cooperative creativity and for yet more elaborate techniques of war fighting. After all, the invention of nuclear weapons is itself a triumph of the human intellect—an ironic manifestation of our capacity for cooperation.

At this point, readers looking to evolution for guidance can be forgiven if they feel confused, even frustrated by the fact that our biological heritage is so ambiguous, or ambivalent. It is certainly worthwhile interrogating the evolutionary background to our predilections, but the answers will likely lead us back to Jean-Paul Sartre's formulation that human beings are "condemned to be free."[22] Advocates for peace might be relieved that we are not biologically obliged to war, or distraught that we are not unalterably predisposed to peace, but we are all stuck with an obligation (if not necessarily a predisposition) to assess our uniquely human situation as honestly as we can.

When dealing with matters of war-proneness versus peaceful capabilities, it would be far better, not only for scientific accuracy but also for social consequence, if we took seriously this pronouncement by one of the premier authorities on human nature, Dr. Seuss: "You have brains in your head. You have feet in your shoes. You can steer yourself any direction you choose."[23]

When it comes to violence and war, evolution's bequeathal to *Homo sapiens* is best symbolized by the two-faced Roman god, Janus, employed in the Gregorian calendar to represent the first month of each year because he looks backward to the year just finished as well as forward to the one that is beginning. Our biological, Janus-faced heritage can similarly predispose human beings toward either violence or peace, depending on the circumstances. One can therefore espouse not only the desirability of peace but also its feasibility. A prerequisite, however, is to free ourselves from the cynical, self-deceiving presumption that our species is biologically doomed to unceasing war.

There is a story, said to be of Cherokee origin, that speaks to this matter, and to our shared human responsibility. A young girl was troubled by a recurring dream in which two wolves fought viciously with each other. When she recounted the dream to her grandfather, a village elder renowned for his wisdom, he explained that there are two wolves struggling inside everyone, one peaceful and the other warlike. At this, the girl was even more upset and asked who wins. Her grandfather's response: "The one you feed."

Old paradigms: (1) Human beings are irrevocably stamped with a biological predisposition to wage war, so we had best get used to it and plan accordingly, or (2) We are inherently benign, benevolent, and peaceful.

New paradigm: We are not biologically doomed to war, although we are inclined to be interpersonally violent on occasion; the war/peace future is in our hands, and isn't written in our genes.

NOTES

1. Some material in this chapter has been modified from an earlier article, D. P. Barash, "Is There a War Instinct," *Aeon,* https://aeon.co/essays/do-human-beings-have-an-instinct-for-waging-war, and from D. P. Barash, "Evolution and Peace: A Janus Connection," *War, Peace and Human Nature,* ed. D. Fry (New York: Oxford University Press, 2015).
2. Raymond Dart, "The Predatory Transition from Ape to Man," *International Anthropological and Linguistic Review,* 1: 201–217 (1953).
3. John Calvin, *Selections from his Writings* (New York: HarperCollins, 2006).
4. Aldous Huxley, *Ape and Essence* (New York: Harper Bros, 1948).
5. Robert Ardrey, *African Genesis* (New York: Atheneum, 1961).
6. J. D. Pruetz et al., "New Evidence on the Tool-Assisted Hunting Exhibited by Chimpanzees (*Pan troglodytes verus*) in a Savannah Habitat at Fongoli, Sénégal," *Royal Society Open Science* 2, no. 4 (2015): 140–507.
7. Dominic Johnson and Bradley A. Thayer, "What Our Primate Relatives Say about War," *The National Interest,* October 6, 2014, http://nationalinterest.org/commentary/what-our-primate-relatives-say-about-war-7996.
8. B. Holmes, "How Warfare Shaped Human Evolution," *New Scientist,* November 12, 2008, https://www.newscientist.com/article/mg20026823-800-how-warfare-shaped-human-evolution/.
9. Bowles, Samuel, "Did warfare among ancestral hunter-gatherers affect the evolution of human social behaviors?" *Science* 324, no. 5932 (2009): 1293–1298.
10. Simon Critchley, Review of John Gray's *The Silence of Animals* (2013), *Los Angeles Review of Books.*

11. Napoleon A. Chagnon, "Life Histories, Blood Revenge, and Warfare in a Tribal Population," *Science* 239, no. 4843 (1988): 985.
12. D. Fry, *Beyond War: The Human Potential for Peace* (Oxford: Oxford University Press, 2007).
13. S. Beckerman et al., "Life Histories, Blood Revenge, and Reproductive Success among the Waorani of Ecuador," *Proceedings of the National Academy of Sciences* 106, no. 20 (2009): 8134–8139.
14. J. Henrich, S. J. Heine, and A. Norenzayan, "Most People Are Not WEIRD," *Nature* 466, no. 7302 (2010): 29.
15. D. Fry and P. Söderberg, "Lethal Aggression in Mobile Forager Bands and Implications for the Origins of War," *Science* 341 (2013): 370–373.
16. S. Pinker, *The Better Angels of Our Nature* (New York: Penguin, 2010).
17. C. Boehm, "Purposive Social Selection and the Evolution of Human Altruism," *Cross-Cultural Research* 42, no. 4 (2008): 319–352.
18. E. O. Wilson, *The Social Conquest of Earth* (New York: Liveright, 2012).
19. E. Sober and D. S. Wilson, *Unto Others: The Evolution and Psychology of Unselfish Behavior* (Cambridge, MA: Harvard University Press, 1999).
20. D. P. Barash, *Homo Mysterious: Evolutionary Puzzles of Human Nature* (New York: Oxford University Press, 2012).
21. B. F. Skinner, *Beyond Freedom and Dignity* (New York: Knopf, 1971).
22. Jean-Paul Sartre, *Existentialism is a Humanism* (New Haven, CT: Yale University Press, 2007).
23. Dr. Seuss, *Oh, the Places You'll Go* (New York: Random House, 1992).

14

About Those Better Angels

ABRAHAM LINCOLN CLOSED his first inaugural address by referring to the "better angels of our nature." If our nature indeed contains some good angels—or at least, inclinations to behave in ways that reflect altruism, kindness, compassion, care for others, and so forth—this seems inexplicable to those people with a naive view of natural selection. It is widely thought that since evolution is a self-centered process (after all, personal gene success—fitness—is what is rewarded), then it shouldn't be capable of generating angelic, self-abnegating behavior. And yet, it is not uncommon for human beings to behave in ways that appear biologically altruistic: enhancing the fitness of someone else, while diminishing that of the altruist.

From here, religious believers such as National Institutes of Health Director Francis Collins go further and claim that the existence of such beneficent traits is prima facie evidence for a divine creator. It is peculiar that a renowned geneticist should be so ignorant of what these days is basic biology, but here it is. "In my view," wrote Dr. Collins in his book *The Language of God*, "DNA sequence alone, even if accompanied by a vast trove of data on biological function, will never explain certain special human attributes, such as the knowledge of the Moral Law and the universal search for God." I deal a bit later with the "universal search for God." Let's first tackle the "Moral Law."[1]

Charles Darwin attempted to do so, and found it downright worrisome. In *The Origin of Species*, he described animal altruism as "one special difficulty, which at first appeared to me insuperable and actually fatal to my whole theory." But fatal it isn't; in fact, with every new discovery in evolutionary biology, we have been obtaining greater confirmation of how Darwin's insights are astoundingly accurate and productive. Not only is altruism no exception, it turns out to be one of the most consistent and confidence-inspiring of all such findings, light years away from anything that requires supernatural invocation.

The reason Darwin saw human morality as especially challenging was because it—which he labeled the "ought"—frequently appeared to go counter to the dictates of natural selection. "Of all the differences between man and the lower animals," he wrote in *The Descent of Man, and Selection in Relation to Sex* (1871), "the moral sense or conscience is by far the most important . . . It is summed up in that short but imperious word *ought*, so full of high significance. It is the most noble of all the attributes of man." Lacking any knowledge of genetics or DNA, Darwin nonetheless proceeded to propose an array of nontheological explanations for human morality, in many cases identifying similar, albeit less clearly developed parallels in animals.

It is now well understood, *pace* Dr. Collins, and extending Darwin, that there are numerous processes whereby natural selection generates what is typically labeled "morality" or "ethics," most of which falls under the rubric "altruism." These include—but are not limited to—kin selection, reciprocity, group selection, third-party effects, parenting behavior, and mutual cooperation. In brief, these mechanisms generate actions that, when observed at the level of individuals and their bodies, appear genuinely altruistic and sometimes even self-sacrificial (hence, highly moral if not downright saintly), but which, at the level of genes, are actually selfish and perfectly within the orbit of natural selection.

MARK TWAIN ONCE noted that it was easy to stop smoking: he had done it hundreds of times! It is similarly easy to explain human morality and altruism: we have many explanations (albeit not hundreds). And unlike Mr. Twain's frustrated attempts to curb his nicotine habit, the abundance of explanations for morality and altruism is not a problem at all, except for those who insist that there must be a single, simple explanation for all things. Instead, biology offers a rich palette of theories, and there is every reason to think that in many cases—perhaps most—more than one operate at the same time.

Best known is the straightforward consequence of gene-based natural selection, the fact that evolution proceeds via differential reproduction—not merely of individuals, but of genes. Although it is easy to identify this perspective as "straightforward," it was anything but that prior to the crucial insights of William D. Hamilton,[2] who helped biologists to understand that genes maximize their evolutionary success (their fitness), by helping to promulgate identical copies of themselves residing in other bodies. Much of the time, these other bodies are known as offspring, which literally explains why living things—including human beings—are inclined to reproduce in the first place. And why they often find themselves doing so even when not inclined.

Hamilton's great insight, however, also provided a crucial new understanding of why many organisms—once again, including human beings—behave altruistically toward other individuals, so long as there is a sufficient probability that identical copies of the genes predisposing toward such behavior exist in those other individuals. This probability is a function of whether individuals are genetic relatives: the closer the relative, the greater the probability that genes in both are identical as a result of their recent common ancestry.

Hamilton introduced the phrase "inclusive fitness," reflecting the fact that natural selection acts on genes so as to maximize their fitness—not merely their reproductive fitness (success in producing offspring and derivatively, the success of their offspring's offspring, and so on) but also recognizing a more inclusive perspective, namely gene success in promoting themselves via the success of collateral relatives (cousins, nieces, nephews, etc.), with the importance of each relative devalued in proportion as it is more distantly related. Acknowledging the significance of kin with respect to inclusive fitness, the biologist John Maynard Smith suggested the term "kin selection."[3] The phrase "indirect fitness" has also been employed, distinguishing the direct, reproductive component of fitness maximization from its indirect, kin-oriented aspect.

Regardless of the terms, the key insight remains unchanged. What appears to be organisms behaving altruistically—benefiting someone else, often at the cost of a decrement in the well-being and fitness of the "behaver"—is actually genes behaving selfishly, benefiting themselves by proxy. This is the gravamen of Richard Dawkins's justly influential book, *The Selfish Gene*,[4] which, as Dawkins himself recognized, could as well have been titled "The Altruistic Gene."

When people help other people, at some cost and/or risk to themselves, we say they are being altruistic, or, more arcanely, that they are acting in accord with a kind of "Moral Law." Cole Porter's song "Let's Do It" points out, "Birds do it, bees do it, even educated fleas do it," concluding, "Let's do it, let's fall in love." Decades of animal behavior research has shown, over and again, that birds do it, bees do it, nearly every living thing that has been carefully studied does it: whether or not they fall in love, they direct beneficence toward genetic relatives. (I don't know about educated fleas, however.) The delightful words of Mr. Porter's song are definitely worth revisiting—for example, "chimpanzees in the zoos do it, some courageous kangaroos do it"—but even readers who refrain from Googling the full lyrics will recognize that human beings do "it": they show favoritism toward their relatives. In short, nepotism is a cross-cultural universal, and although it isn't usually considered especially moral, its connection to altruism via kin selection is stunningly clear.

What about altruism or a sense of moral responsibility toward individuals with whom you do *not* have genes in common based on a recent common ancestor? For a powerful additional theory, we turn once more to the work of Robert Trivers, whom we met earlier when discussing parent–offspring conflict. Trivers published a landmark paper in 1971 titled "The Evolution of Reciprocal Altruism,"[5] in which he presented a model for understanding altruism between individuals who are not only unrelated but also—in theory at least—even different species. The basic idea is common sense: you scratch my back, and I'll scratch yours. Imagine that your back could use some scratching, but you can't do it yourself, and you have no obliging genetic relatives around. You might solicit someone else to do the deed with the expectation that at some time in the future, the amiable scratcher will herself feel an itch, and you will reciprocate. If so, then both of you would end up better off. (To my mind, since both the initiator and recipient experience higher fitness as a result of their interaction, no one is being altruistic—which is normally taken to mean an exchange in which the altruist loses immediate fitness while the recipient gains—hence, "reciprocal altruism" is a misnomer. I prefer to designate the encounter "reciprocity," without the "altruism," although this is quibbling.)

Trivers recognized that there is a problem with reciprocity; namely, what if the recipient doesn't reciprocate? What if, having obtained some benefit, the recipient acts selfishly and refuses to respond in kind when the initial scratcher feels an itch? That individual had expended time and energy helping someone else, someone who is unrelated and therefore, whose benefit did not in itself rebound automatically to aid the fitness of the helpful scratcher. If that recipient later "cheats" and doesn't pay back her behavioral debt, then the cheater would end up ahead and the initial helper, worse off; that is, she would be a genuine altruist. Natural selection would therefore work against taking the risk of being helpful to a nonrelative in the first place.

The situation, as Trivers pointed out, is very similar to what game theorists call the Prisoner's Dilemma, a situation in which a player is penalized for being nice or cooperative, if the other player takes advantage of the situation to be mean or uncooperative—game theorists use the term "defect." In such a case, defectors do better than cooperators, and so both players end up being nasty defectors, fearful of being suckered by the other, while also tempted to take advantage of the other player in case she might play nice. It's a dilemma because in such cases, both participants would be better off cooperating, but because of the painful logic of the game, they each end up with the punishing payoff of mutual defection or selfishness. In our hypothetical itch-scratching example, everyone would be most fit if they got their itches scratched, but they are at risk of ending up with an itchy and unscratched back—worse yet, after having expended time, energy, and perhaps risk while helping an ingrate.

There have been many analyses of the Prisoner's Dilemma, out of which one strategy—known as tit-for-tat—sometimes emerges as a way out. This isn't the place to explore the complexities of the Prisoner's Dilemma. Suffice it for our purposes to note that backscratching might be one case in which reciprocity could evolve, especially because the cost of the initial act is low (it usually doesn't take much time, or energy, or impose much risk to scratch someone else's back), and the probability is that reciprocity will be similarly low-cost, and therefore likely to happen. For reciprocity to evolve, it is also necessary that the individuals involved are prone to maintaining an interactive relationship over time, so that they are in proximity, and hence, available to return the favor when called on. It also helps if the participants are smart enough to recognize each other as individuals, so that the one owing a return favor can identify to whom it is due, and for the earlier beneficiary to be able to recognize the debtor and if need be, remind him or her of the obligation. Moreover, it would be better yet if the individuals in question also participated in a social system that pressured cheaters and threatened to punish them—most likely, by refusing to engage in a potentially reciprocal relationship in the future.

Very few bona fide cases of reciprocity have been found among free-living animals. One of these involves, of all animals, vampire bats.[6] These creatures have small bodies and a high metabolic rate; adding to the difficulty of being a vampire bat, their nocturnal foraging is often unsuccessful. If they fail to obtain a blood meal more than two days in a row, they are in danger of starving. Vampire bats roost in social groups, and it is apparent from their engorged abdomens who has fed successfully when they return after a night's foraging. Those who didn't get to obtain a blood meal from a sleeping cow or horse adopt a begging posture toward those who did, and the latter are likely to regurgitate some of their bounty. Not only that, but when the tables are turned—as is often the case, since success and failure appear to be largely random—a previous recipient is especially likely to reciprocate and share food with its earlier benefactor. It may also be noteworthy that despite their unsavory reputation, vampire bats have exceptionally large brains for their body size, an adaptation that might be a result of selection favoring individuals who were able to identify debtors and demand that they reciprocate.

What about people? We certainly engage tit-for-tat reciprocation, especially between nonrelatives. Although this is rarely acknowledged, we also maintain close mental accounts covering whom we helped and whether they reciprocated, as in the case of dinner invitations and even extending to such small details as Christmas cards. Friendship itself isn't normally seen as motivated by a hard-headed calculus of "What have I done for you, and what have you, in turn, done for me lately . . . or ever?" Yet it may well turn out that mutual aid is "what friends are for," and as mutuality goes, so goes the friendship. Of course, it is also possible—and in many cases unavoidable—that networks of reciprocity are overlain on an even denser social and moral fabric woven by shared genes. One doesn't preclude the other.

Another possible avenue to altruism and mutually agreed morality is group selection. We've already briefly considered this scientifically controversial mechanism when looking at humanity's presumed war-proneness. With respect to the topic of this chapter, the idea is that those better angels of our nature might have evolved—at least in part—because groups composed of moral, altruistic, angelic individuals could have enjoyed greater success than groups composed of selfish, uncooperative, defection-prone SOBs. As with reciprocity, however, the worm in the blossom of altruism via group selection is the problem of cheaters, or "free riders." These are individuals who take advantage of their more generous colleagues, and benefit by their generous, angelic natures, but persist in maintaining their selfish ways—who take but don't give. It is nonetheless at least possible that group selection has contributed to the evolution of human beneficence; moreover, given our big brains, the persistence of social groups (at least before very recent times), as well as our penchant for moralistic aggression toward noncooperators, it appears that *Homo sapiens* might be the species most prone to reciprocity, sharing and thus, what is widely designated interpersonal morality.

Maybe even more than vampire bats.

KIN SELECTION, RECIPROCITY, and group selection don't exhaust the possible biological causes of human moral behavior. There is also "indirect reciprocity," aka third-party effects. The basic idea is that moral behavior, including altruism and even self-sacrificial heroism, can have been favored by natural selection if "good" individuals are perceived as sufficiently praiseworthy that the cost incurred by their beneficent act is outweighed on balance by the positive returns an enhanced reputation could provide: this might include but not be limited to greater mating opportunities, sharing of resources, benefits accruing to their offspring and other relatives, and so forth.

Altruism and morality could also derive partly from a near-universal human trait: parenting. After all, being a good or just an adequate parent requires a degree of empathic, helpful responding to the needs of someone who is much smaller, younger, and needier than one's self: recognizing indications of distress, providing care, and assistance of all sorts.[7] A number of studies have confirmed, in addition, that a compassionate mind-set has physiologically beneficial consequences for those so disposed.[8] It is not at all unlikely, therefore, that altruistic care for others evolved as an extension of parental care, which not only feels good but is good for one's evolutionary success. In short, the underlying neural, hormonal, and psychophysiological mechanisms that were almost certainly initiated in the context of selection for taking care of one's offspring could readily have been deployed in taking care of and caring about others.[9]

Of course, for natural selection to have promoted this transition, some sort of ultimate adaptive value would have been required, and here kin selection, reciprocity, and group selection might well have sufficed. If so, then just as our arboreal primate heritage set the stage for human binocular vision and our fondness for sweet fruits, parenting could have supplied the proximate physiological precursors for beneficent care-taking and concern for the well-being of others. In that regard, it probably isn't coincidental that human beings are not only unusual in the degree of altruism that we manifest but also extraordinary in the helplessness of our newborns and the degree of prolonged parental care that they require.

As part of his effort to identify the sources of human altruism and morality, Darwin had wondered specifically about heroic self-sacrifice (remember, he lived a full century before the various evolutionary routes to altruism described above had been identified). He could not understand how the "offspring of the more sympathetic and benevolent parents, or of those which were the most faithful to their comrades, would be reared in greater number than the children of selfish and treacherous parents," not to mention the seemingly evident fact that a hero "would often leave no offspring to inherit his noble nature."

It appears that society, too, has had similar worries. Aside from suppressing the kind of "noble nature" that seems likely to benefit others in the community, selfish tendencies can threaten the smooth functioning of any social group. One result has been societal condemnation of such selfishness, combined with active efforts to promote altruism, rewarding the most outwardly benevolent and condemning those who are so self-serving as to be troublesome to everyone else.

Friedrich Nietzsche, who criticized Christianity as a "slave morality" that rewarded the weak (read here, the altruists) while punishing the strong (i.e., the selfish) maintained that society—and not just Christianity—encourages self-sacrifice because an other-regarding sucker is an asset to those others. In *The Gay Science*, Nietzsche wrote that

> Virtues (such as industriousness, obedience, chastity, piety, justness) are mostly injurious to their possessors . . .. If you possess a virtue . . . you are its victim! But that is precisely why your neighbor praises your virtue. Praise of the selfless, sacrificing, virtuous . . . is in any event not a produce of the spirit of selflessness! One's "neighbor" praises selflessness because he derives advantage from it!

Two centuries earlier, the French moralist Jean de la Bruyere observed that as far as society is concerned,

> That man is good who does good to others. If he suffers on account of the good he does, he is very good; if he suffers at the hands of those to whom he has done good, then his goodness is so great that it could be enhanced only by greater suffering, and if he should die at their hands, his virtue can go no further; it is heroic. It is perfect.[10]

But one need not be a cynic like de la Bruyere or Nietzsche to recognize the pressure that is typically exerted to induce people to be, as Bertholt Brecht put it in *The Threepenny Opera*, "more kindly than they are." Freud identified the superego—analogous to the conscience, and imposed particularly by one's parents—as a major driver of human behavior, similar to his argument in *Civilization and Its Discontents* that civilization itself is built on the suppression of our id-like instincts. In other words, we are enjoined by our parents, by social norms, by religious teaching, and by the threat of public shaming to be good citizens—which is to say, practitioners of an acceptable form of interpersonal morality.

On the one hand, such manipulation wouldn't be necessary if people were always inclined to behave—"naturally"—as saints rather than sinners. But at the same time, given the proethical arm-twisting that appears characteristic of all human groups, added to the biological pressures generated by kin selection, reciprocity, third-party effects, and possibly group selection as well, there is no lack of momentum encouraging people to be good.

Finally, there is yet another potential wellspring of altruism and morality that makes no claims of supernatural indulgence. Darwinian evolution is typically seen as synonymous with dog-eat-dog, no-holds-barred competition. In that regard, the poet Robinson Jeffers is on target in his poem "The Bloody Sire," where he notes the role of selection in generating the biological equivalent of arms races in sharpening the capacities of predator and prey alike:

> What but the wolf's tooth whittled so fine the fleet limbs of the antelope?
> What but fear winged the birds,
> And hunger jeweled with such eyes the great goshawk's head?

Within-species competition (especially but not uniquely among males) has also played a role in shaping much of a species' anatomy as well as its behavior. At the same time, there are many circumstances in which selection also rewards a degree of social cooperation, as with collaborative hunting, antipredator warning calls, the sharing of information about food and water sources, and even huddling together to keep warm. The mathematical biologist Martin Nowak suggested that in addition to "struggle," we should also recognize the widespread "huddle for survival."[11]

Ornithologists have long known that cooperative breeding systems are most frequent among birds that occupy challenging environments. A study of the life history and movement patterns of four thousand different bird species concluded that the effect worked both ways: those species engaging in cooperative breeding were more likely to successfully colonize difficult environments.[12] Part of the human story is of a species that migrated out of Africa, to colonize a wider range of environments than any other vertebrate—from the Arctic to the tropics, high altitude to deep valleys, grassland to forest, sea-shore to desert—a biogeographic accomplishment that would probably have been impossible if our ancestors were not only restless but also good at cooperating, and not just when it comes to childrearing. It takes a village not only to rear a child but also to flourish in a demanding world. Not surprisingly, it has been suggested that human cooperation—in a variety of contexts—may have been fundamental to the evolution of altruism generally.[13]

Serious theologians have little fondness for the "God of the gaps" formulation, the argument that God can and even should be invoked to explain whatever science cannot. Their problem is that as science advances and continues to fill in the gaps in our knowledge, those gaps—and thus, the space for god—become ever smaller. I very much doubt that Francis Collins had honestly struggled to find scientific explanations for how human beings might have come to our "knowledge of the Moral Law" before suggesting that only God can explain it. Had he consulted the biological literature, Dr. Collins would quickly have discovered that rather than an explanatory gap, there is—à la Mark Twain—a surfeit of perfectly good, materialistic scientific explanations. Moreover, unlike religious fundamentalism, which typically emphasizes one literal and unquestionable way of seeing the world, a scientific approach recognizes that truth is often nuanced and multifactorial.

I hope readers are not put off by the multiplicity of explanations for human altruism. Explanations for natural phenomena are in fact more powerful, and even more likely to be true, if they are multivalent than if they are unitary. Insofar as many roads lead to Rome, such a multiplicity makes it more likely that a traveler will eventually get there. Bon voyage!

IT IS WIDELY reported that when in 1802 Napoleon met with the physicist and mathematician Pierre Laplace, the French emperor noted that in Laplace's detailed book on celestial mechanics, there was no mention of the universe's "divine creator," to which Laplace replied, "I had no need of that hypothesis."

When it comes to explaining the "better angels of our nature," neither do we.

Old paradigm: The human penchant for altruism, beneficence, caring for others, and moral sensibility could not have evolved via a brute mechanical process of natural selection; hence, it is evidence for god.

New paradigm: There are many plausible biological explanations for these traits, which are not uniquely human, and which do not require—or even suggest—divine intervention.

NOTES

1. Francis Collins, *The Language of God* (New York: Simon and Schuster, 2006).
2. William D. Hamilton, "The Genetical Evolution of Social Behaviour. II," *Journal of Theoretical Biology* 7, no. 1 (1964): 17–52.
3. John Maynard Smith, "Kin Selection and Group Selection," *Nature* 201, no. 1 (1964): 1145–1147.
4. Richard Dawkins, *The Selfish Gene* (Oxford: Oxford University Press, 1976).
5. Robert L. Trivers, "The Evolution of Reciprocal Altruism," *Quarterly Review of Biology* 46, no. 1 (1971): 35–57.
6. G. S. Wilkinson, "Reciprocal Food Sharing in the Vampire Bat," *Nature* 308, no. 5955 (1984): 181–184.

7. S. D. Preston, "The Origins of Altruism in Offspring Care," *Psychological Bulletin* 139 (2013): 1305–1341.

8. O. M. Klimecki et al., "Differential Pattern of Functional Brain Plasticity after Compassion and Empathy Training," *Social Cognitive and Affective Neuroscience* 9 (2014): 873–879; A. Kogan et al., "Vagal Activity Is Quadratically Related to Prosocial Traits, Prosocial Emotions, and Observer Perceptions of Prosociality," *Journal of Personality and Social Psychology* 107 (2014): 1051–1063.

9. G. Loewenstein and D. A. Small, "The Scarecrow and the Tin Man: The Vicissitudes of Human Sympathy and Caring," *Review of General Psychology* 11 (2007): 112–126.

10. Jean de La Bruyère. *The "Characters" of Jean de La Bruyère* (London: John C. Nimmo, 1885).

11. Martin A. Nowak, "Why We Help," *Scientific American* 307, no. 1 (2012): 34–39.

12. C. K. Cornwallis et al., "Cooperation Facilitates the Colonization of Harsh Environments," *Nature Ecology and Evolution* 1 (2017): p.57.

13. F. Warneken and M. Tomasello, "The Roots of Human Altruism," *British Journal of Psychology* 100 (2009): 455–471.

15

Who's in Charge?

THE ANSWER SEEMS obvious: each of us controls what goes on inside his or her head. The heroic closing lines of "Invictus" by the Victorian poet William Ernest Henley express this: "I am the master of my fate; I am the captain of my soul."

Not so fast, however. Aside from the fact that "inside" and "outside" aren't so clearly distinguished (see chapter 8), there are many claimants to being master and captain or—more intellectually troublesome—reason to suspect that no one is minding the store, guiding the ship, calling the shots, giving the orders, and so forth. Or at least, that the hand on the tiller isn't the same from one action to the next. A popular expression these days relates to whether a particular activity is "in my wheelhouse," meaning within someone's zone of competence. A popular realization these days, among biologists in particular, is that whereas ideas and prospective actions are indeed more or less susceptible to individual competence, no one consistently resides inside one's own wheelhouse.[1]

This leads inevitably to the ancient philosophical conundrum of free will, one that has been given new impetus by advances in neurobiology. We know that, as Francis Crick put it in *The Astonishing Hypothesis*, "You, your joys and your sorrows, your memories and your ambitions, your sense of personal identity and free will, are in fact no more than the behavior of a vast assembly of nerve cells and their associated molecules."[2] The problem here is that we all operate under the illusion that we possess free will, and yet, neurons function as a result of such physical processes as electrochemical and osmotic gradients acting across cell membranes and on the physical structures of axons, dendrites, and synapses. If all these happenings are material events responding "automatically" to preceding conditions, where is our free will? Where is the little green man or woman inside our head who presumably acts on his or her "own"?

Answer: Nowhere. There is no such entity; hence, we are stuck with a deep paradox: any rigorously scientific view of free will must conclude that it is an illusion, a troublesome but unavoidable conclusion that is ultimately traceable to the fact that our neurons—no less than the rest of our "selves"—are connected to and exquisitely sensitive to physical events happening both "outside" and "inside" our heads. And yet, we all live our lives under the illusion of free will, applying it to ourselves as well as to others in our lives. This suggests the existence of yet another widespread illusion, perhaps another paradigm that we must lose: that we are consistent in our beliefs!

For our purposes, the inevitable scientific refutation of free will coupled as it is with the near-universal certainty that we are nonetheless independent and autonomous actors, is less paradigm-busting than it might seem at first glance. The debate over free will versus determinism long predated modern insights from neurobiology.[a] Rather, the paradigm that has been actively crumbling is one involving the ostensibly unitary mental actor within everyone's head.

Thus, closely related to free will—and somewhat more susceptible to consistent insights—is the phenomenon of "host manipulation," a new and exciting scientific frontier that requires modifying our expectation of being independent actors who perform in our own interest. Consider the disconcerting fact that there are many more freeloaders than free-livers, many more parasites and pathogens than individual, stand-alone organisms. After all, pretty much every multicellular animal is home to numerous fellow travelers, and—this is the point—each of these creatures has, in a sense, its own agenda. The invertebrate biologist Ralph Buchsbaum, considering just one group of worms, suggested,

> if all the matter in the universe except the nematodes were swept away, our world would still be dimly recognizable . . . Trees would still stand in ghostly rows representing our streets and highways. The location of the various plants and animals would still be decipherable, and, had we sufficient knowledge, in many cases even their species could be determined by an examination of their erstwhile nematode parasites.[3]

Put another way, if a biologist were to answer Frank Zappa's acid-rock question "Suzy Creamcheese, honey, what's got into you?" the response might well be: "A whole lot of other living things."

What difference does this make? For many of us supposedly "free-living" creatures, quite a lot. Providing room and board to other life forms doesn't only compromise one's nutritional status (not to mention peace of mind), it often reduces freedom of action, too. As unwitting hosts, organisms are manipulated by their occupants in many different ways.[4] Case in point: the tapeworm *Echinococcus multilocularus* causes the mouse in which it resides to become obese and sluggish, making it easy pickings for predators, notably foxes which—not coincidentally—constitute the next phase in the tapeworm's life cycle. Those the gods intend to bring low, according to the Greeks, they first make proud. Those tapeworms intending to migrate from mouse to fox do so by first making "their" mouse fat and sluggish, and thus, fox food.

---

[a] When asked if he believed in free will, Isaac Bashevis Singer replied, "I have no choice."

Sometimes the process is more bizarre. For example, the life cycle of trematode worms known as *Dicrocoelium dentriticum* involves doing time inside a snail, after which the worms are shed inside slime balls, which are then eaten by ants. Now these worms must go from ants to their next stopover: sheep. This requires some fancy manipulation of the unwitting ants: ensconced within their "host" ant, some of the worms become downright antsy. They migrate to the ant's brain, where they rewire their host's neurons and hijack its behavior. The manipulated ant, acting with zombie-like fidelity to *D. dentriticum*'s demands, then climbs to the top of a blade of grass and clamps down with its jaws, whereupon it waits patiently and conspicuously until it is consumed by (no surprise!), a grazing sheep. Thus transported to its desired happy breeding ground deep inside sheep bowels, the worm turns, or rather, releases its eggs, which depart along with a healthy helping of sheep poop, only to be consumed once more by snails and then by ants; repeat ad infinitum. It's a distressingly frequent story—distressing, at least, to those committed to "autonomy."

Another example from the natural world, as unappetizing as it is important: plague, the notorious Black Death, is caused by bacteria carried by fleas, which, in turn, live mostly on rats. Rat fleas sup cheerfully on rat blood, but will happily gorge on people, too, and when they are infected with the plague bacillus, they spread the illness from rat to human. The important point for our purposes is that once they are infected with plague, disease-ridden fleas are especially enthusiastic diners, because the plague bacillus multiplies within flea stomachs, diabolically rendering the tiny insects incapable of satisfying their growing hunger. Not only are these fleas especially voracious in their frustration, but also, because bacteria are cramming its own belly, an infected flea vomits blood back into the wound it has just made, introducing yet more plague bacilli into their victim. A desperately hungry, plague-promoting flea, if asked, might well claim, "the devil made me do it," but in fact, it is the handiwork of *Pasteurellis pestis*.

Not that a plague bacterium—any more than a mouse-dwelling tapeworm or ant-hijacking "brain-worm"—knows what it is doing when it reorders the inclinations of its host. Rather, a long evolutionary history has arranged things so that the manipulators have inherited the earth and that they do so by depriving their victims of autonomy. ("So, naturalists observe, a flea has smaller fleas that on him prey," wrote Jonathan Swift. "And these have smaller still to bite 'em. And so proceed ad infinitum.")

The ways of natural selection are devious and deep, embracing not only would-be manipulators but also their intended victims. Take coughing, or sneezing, or even—since we have already broached some indelicate matters—diarrhea.

When people get sick, they often cough and sneeze. Indeed, aside from feeling crummy or possibly running a fever, coughing and sneezing are important ways we identify being ill in the first place. It may be beneficial for an infected person to cough up and sneeze out some of their tiny organismic invaders, although it isn't beneficial for others nearby. This, in turn, leads to an interesting possibility: what if coughing and sneezing aren't merely symptoms of disease and indicators of the body's efforts to cleanse itself but also—even primarily—a manipulation of us, the "host," by, say, influenza virus? Mice, as we have seen, can be fattened and made less cat-wary, just as ants can be made grass-blade-besotted. Perhaps people are similarly made to cough and sneeze by organisms whose transmission is thereby benefited.

As to diarrhea, what a great way for an intestine dwelling pathogen to spread itself to new victims. Cholera, for example, causes terrible (and potentially deadly) diarrhea. As with a flu victim's sneezing and coughing, perhaps it benefits a cholera sufferer to expel some of the cholera-causing critter, *Vibrio cholerae*. But it also benefits *V. cholerae*. Just as Lenin urged us to ask "who, whom?" with regard to socioeconomic interactions— Who benefits at the expense of whom?—an evolutionary perspective urges on us the wisdom of asking a similar question. Who benefits when a cholera victim "shits his guts out," whether or not she dies? Answer: the cholera bacillus.

This most dramatic symptom of cholera is caused by a toxin produced by the bacillus, making the host's intestines permeable to water, which gushes into the gut in vast quantities, after which more than 100 billion *V. cholerae* per liter of effluent sluices out of the victim's body, whereupon, if conditions are less than hygienic, they can infect new victims. At the same time, *V. cholerae* benefit doubly from their victim's agony, since the colonic flood that spreads the pathogen also washes out much of the normal, native bacterial intestinal flora, leaving a comparatively competitor-free environment in which those *V. cholerae* that are left behind can flourish. Diarrhea, then, isn't just a symptom of cholera; it shows every sign of being a successful manipulation of *Homo sapiens*, by and for the bacteria.

Intriguing as these tales of pathology may be, it is too easy to shrug them off when it comes to the daily, undiseased lives most of us experience. After all, aside from sneezing, coughing, or pooping, our actions are, we like to insist, ours and ours alone, if only because we are acting on our own volition and not for the benefit of some parasitic or pathogenic occupying army. So, when we fall in love, we do so for ourselves, not at the behest of a romance-addled tapeworm. When we help a friend, we aren't being manipulated by an altruistic bacterium. If we eat when hungry, sleep when tired, scratch an itch, or write a poem, we aren't knuckling under to the needs of our nematodes. But in fact, it isn't that simple. There is every reason to suppose that we aren't nearly as much "in control" as we assume.

Think about having a child. Not how it feels to become a parent, nor what it costs, or the social, family, personal, or cultural factors involved. Rather, think about it as Lenin suggested (who, whom?), or as a modern-day Darwinian might: Who—or rather, what—benefits from reproduction? Return to the ruling mantra of revolutionary biology: It's the genes, stupid. They're the beneficiaries of baby making, the reason for reproducing, just as they are the reason for kin-based altruism (chapter 14). As modern evolutionary biologists increasingly recognize, bodies—more to the point, babies—are our genes' way of projecting themselves into the future. Or as Richard Dawkins has so effectively argued, bodies are temporary, ephemeral, short-lived survival vehicles for those selfishly manipulative genes, which are the only entities that persist over evolutionary time. No matter how much money, time, or effort is lavished on them, regardless of how much they are exercised, pampered, or monitored for bad cholesterol, bodies don't have much of a future. In the scheme of things, they are as ephemeral as a spring day, a flower's petal, a gust of wind. Bodies go the way of all flesh: ashes to ashes and dust to dust, molecule to molecule and atom to atom.

AND SO, WE return to some key questions: Who is in charge? Who is calling and who is heeding? The biologically informed answer—albeit not one that makes the heart sing—is not all that different from those alarming rat/tapeworm, ant/trematode, flea/bacteria relationships, only this time its genes/body. Unlike the cases of parasites or pathogens, when it comes to genes manipulating "their" bodies, the situation seems less dire to contemplate, if only because it is less a matter of demonic possession or an unfriendly takeover than of *our* genes, *our* selves. The problem, however, is that those presumably personal genes aren't any more hesitant about manipulating ourselves than is a brain worm hijacking an ant.

Take a seemingly more benign behavior, indeed, one that is highly esteemed: altruism. As we have already seen, this is a favorite of evolutionary biologists, because superficially, every altruistic act is a paradox, a mystery in need of solving. Natural selection should squash any genetically mediated tendency to confer benefits on someone else while disadvantaging the altruist. Such genes ought, therefore, to disappear from the gene pool, to be replaced by their more selfish counterparts; altruism should therefore be extremely rare, popping up occasionally as a result of a random mutation but not persisting. To a large extent, however, the paradox of altruism has been resolved by the recognition that "selfish genes"—in the sense of Dawkins and others—can promote themselves (rather, identical copies of themselves), by conferring benefits on genetic relatives who are likely to carry copies of the genes in question.

By this process, known as "kin selection," behavior that appears altruistic at the level of bodies is revealed to be selfish at the level of genes. Nepotism is natural. (So, by the way, is our old friend the cholera-causing bacillus; natural and "good" aren't necessarily the same.) When someone favors a genetic relative, who, then, is doing the favoring: the nepotist, or the nepotist's genes? Just as sneezing may well be a successful manipulation of "us" (*Homo sapiens*) by "them" (viruses), what about altruism as another successful manipulation of "us," this time by our own "altruism genes"? Admirable as altruism may be, it is therefore, in a sense, yet another form of manipulation, with the manipulated victim (the altruist) acting at the behest of some of his or her own genes. The fact that this typically happens when there is a sufficient probability that the beneficiary contains identical copies of those genes is what makes the act—seemingly altruistic from the perspective of the bodies involved—actually selfish at the level of genes. After all, just as the brain worm gains by orchestrating the actions of an ant, altruism genes stand to gain when we are nice to cousin Sarah, never mind that such niceness is by definition costly for the helper.

Who is in charge, therefore? Us or our genes?

All this may seem a bit naive, in several ways. For one, without our genes there would be no "us." For another, even if we grant some sort of distinction (e.g., between genotype and phenotype, between DNA and body) biologists know that genes don't order their bodies around. No characteristic of any living thing emerges full-grown from the coils of DNA, like Athena leaping out of the forehead of Zeus. As we have already emphasized, every discernible trait of any living thing—including behavior—results from a complex interaction of genetic potential and experience, learning as well as instinct, nurture inextricably combined with nature. Life is subject to genetic influence, not determinism.

But does this really resolve the problem? Let's say that a brain-worm-carrying ant still possesses some free will. And that a trematode-carrying mouse has even a bit more, insofar as mice have more complex brains and thus, presumably, more independence of action. What if a mouse's behavior, compared to that of an ant, is influenced by its fellow travelers, be they pathogen, parasite, or gene, and is thus less determined than that of an ant? Wouldn't even "influence" be enough to cast doubt on that mouse's agency, its independence of action? And (here comes at least one Big Question), why should human agency or free will be any less suspect? Even if we are manipulated just a tiny bit by our genes, isn't that enough to raise once again that disconcerting question: Who's in charge here?

Maybe it doesn't matter. Or, put differently, maybe there is no one in charge. That is, no one distinguishable from everyone and everything else. If so, then this is because the "environment" is no more outside us than inside, part tapeworm, part bacterium, part genes, and—once again—no independent, self-serving, order-issuing homunculus (chapter 8). "We" are manipulated by, no less than manipulators of, the rest of life inside and all around us. And so, the answer to "Who's in charge here?" depends on what the meaning of "who" is. Who is left after the parasites and pathogens and removed? Or after "you" are separated from your genes?

Biologists often confront other, comparable issues of self *versus* not-self, notably when it comes to immune responses and the body's natural tendency to reject "foreign" tissue. This, in turn, requires that in the case of organ transplants, various drugs must be used to eliminate—or at least to blunt—such rejection responses. Interesting problems also arise when it comes to defining the literal limits of individuality. What, for example, should we make of clones—artificial as promised (or threatened) by advances in biotechnology, as well as natural, such as we find in the case of aspen groves, which, although they appear to consist of individual trees, are actually one "individual," connected via underground root systems?

Then there are those unicellular organisms that reproduce by simply dividing, as with amoebas: we generally consider that once the cell membrane that had comprised one amoeba has pinched off, thereby leaving two "daughter cells," that the amoeba has lost its individuality. Yet, when cells that eventually make up a human body reproduce in precisely the same way—via mitosis—they remain somehow part of the larger whole that we insist (somewhat perversely and unbuddhistically) on labeling one distinct and separate "individual."

The situation is more fraught still. Consider, for example, the curious fact that within each healthy human being there are about 100 trillion cells (most of them occupying the small and large intestine) that are literally "foreign" in that they are microbes with a completely different genome from that of the body they inhabit. Yet they aren't parasites or pathogens; rather, they are crucial to health, involved not only in digesting nutrients but in generating appropriate immune responses, producing a variety of necessary enzymes, neurotransmitters, and various biochemical substances while doubtless serving other functions not yet identified.

Given all this perplexity, let's leave the last words to a modern icon of organic, oceanic wisdom, SpongeBob Squarepants, a make-believe character recognizable to many children and more than a few adults. Mr. Squarepants is a cheerful, talkative—although admittedly, somewhat cartoonish—fellow of the phylum Porifera, who, according to his theme song, "lives in a pineapple under the sea . . . Absorbent and yellow and porous is he." I don't know about the pineapple or the yellow, but absorbent and porous are we, too.

Old paradigm: Aside from obvious constraints, each of us is in control of his or her life, if not an "army of one," at least the chief operating officer of our own central intelligence agency.

New paradigm: Everyone is shot through with a diverse array of other organisms as well as other entities, each exercising influence on the levers of "control," such that either no one is in control or everyone is . . . whatever that means!

NOTES

1. Some material in this chapter has been modified from chapter 2 of my book *Buddhist Biology*.
2. Francis Crick, *The Astonishing Hypothesis* (New York: Touchstone, 1995).
3. Ralph Buchsbaum, *Animals without Backbones* (Chicago: University of Chicago Press, 1987).
4. D. P. Hughes, J. Brodeur, and F. Thomas, *Host Manipulation by Parasites* (New York: Oxford University Press, 2012).

16

The Paradox of Power

A KEY MESSAGE of this book has been anthrodiminution, the often painful, but objectively unavoidable recognition that we're not as special as many people like to think. We are neither literally nor metaphorically central to the cosmos, and, moreover, we aren't all that extraordinary when it comes to our own "human nature," which is a deeply material consequence of evolution by natural selection, shared in various ways with other animals. And yet, we are certainly special in some respects, many of them quite admirable, not least our capacity for creative accomplishment including scientific insights and discoveries, achievements in the arts and humanities, and the elaboration of ethical and moral guidelines, including, on occasion, acts of surpassing courage, generosity, and kindness. At the same time, we are special in an immensely dangerous respect: our capacity for destruction, apparent across many dimensions. Even as we are called to acknowledge that we lack any trace of metaphysical uniqueness, we are also obliged to recognize our immense power to influence the rest of the planet—and the responsibility that comes with it.

In the past, most problems faced by our species were caused by external events, such as floods, droughts, earthquakes, hurricanes, tornadoes, volcanic eruptions, and so forth—as well as challenges posed by other species, notably those we sought to eat and that endeavored to eat us. Now, we are especially confronted with troubles of our own making.

Anthropogenic climate change is high on the list of our self-made threats, but even its horrors are exceeded by that of nuclear holocaust, a danger that seemed for a time to have retreated with the end of the Cold War, but which is back once again, at least as devastating as ever, and perhaps more probable. Moreover, in view of our connectedness to everything else in the natural world, "we are destroying not only our home, which is dreadful enough, but also a fundamental part of ourselves."[1]

We got to this pass by a kind of Faustian bargain, in which we gained power but at a terrible cost. One way to envision this process is to note that compared to most other animal species, the human body is quite unimpressive, at least when it comes to our inherent armament: we lack claws, fangs, horns, or antlers. Our teeth, in fact, are absurdly small and our jaws are ridiculously receding. We cannot fly, and swim only after considerable training. We can run, but not nearly as quickly as either our potential prey or—more to the point—our prehistoric predators. We can't swing from the trees like our ape cousins, and without the benefit of external clothing and the intervention of cooking, we can neither survive in most environments nor even nourish ourselves. And yet, look around: the human species has not only survived, but prospered—to the extent that our excessive numbers contribute substantially to the many existential threats that humanity faces.

How have we managed this feat? Unlike Faust, we didn't give up our soul, not only because we never had one but also because there was no Mephistopheles. Our peculiarly human power grab did not occur because we gave something up or somehow made a bad deal, but because we managed to achieve what has been in many ways a very good deal: mixing together two seemingly disparate processes within ourselves, phenomena that—like everything else about us—are entirely natural, but unique in their potency when combined. The result has been immense power and success, but also danger that is equally great.

On the one hand, there is our biological evolution, a relatively slow-moving organic process that can never proceed more rapidly than one generation at a time, and that nearly always requires an enormous number of generations for any appreciable effect to arise. Biological evolution bequeathed us a notably large and creative brain, with a mind to match. On the other hand is cultural evolution, a process that is, by contrast, extraordinary in its speed. Whereas biological evolution is Darwinian, moving by the gradual substitution and accumulation of genes, cultural evolution is Lamarckian, powered by a nongenetic "inheritance" of acquired characteristics. During a single generation, people have selectively picked up, discarded, manipulated, and transmitted cultural, social, and technological innovations that have become almost entirely independent of any biological moorings.

In their early stages, these two processes would likely have been mutually reinforcing. Thus, our capacity for culture is itself our most potent biological adaptation, enabling a population of weak-bodied Pleistocene primates to gain ascendancy over most of the world. At the same time, as cultural evolution proceeded and we endowed ourselves with language, impressive cognition, complex social organization, and increasingly powerful technology, the ability to employ these skills in order to master our immediate environment almost certainly reflected back onto our biological evolution. It seems inevitable that as these cultural skills and processes developed and provided leverage over the material and natural world—not to mention over other human beings, less adroit at these things—natural selection favored those individuals maximally able to take advantage of such traits. If so, then individuals with larger and more effective brains, who were therefore more adroit at employing language plus cognitive and problem-solving skills and better able to manipulate complex social organization and to develop and wield increasingly elaborate and powerful technology would have literally experienced greater reproductive success, resulting in human beings whose biological traits "played nicely" with their cultural capacities and innovations.

Up to a point, therefore, our biological and cultural evolution would have been mutually reinforcing. But we are now well past that point.

There is, after all, no reason for our biological and cultural evolution to proceed in lockstep—and many reasons for them to have become disconnected. As already noted, biological evolution is chained to genetic change (indeed, that's what biological evolution *is*), which is necessarily slow, certainly no faster than reproduction itself. For *Homo sapiens*, a single generation requires something like a minimum of two decades, and since any significant evolutionary change requires hundreds, thousands, and even tens of thousands of such generations, we are talking about thousands, tens of thousands, hundreds of thousands, even millions of years.

Cultural evolution is entirely different. Although it fundamentally depends on a biological infrastructure (it takes brains to make a fire, a computer, a functioning language, a hydrogen bomb), cultural "mutations" can arise and be adopted, modified, and passed along within days, without waiting for the passing of even a single biological generation. I am writing this on a computer, having purchased my first "word processor" in the mid-1980s. Before that, I wrote on an IBM Selectric typewriter, and before that, on a Smith Corona manual. All within just one lifetime. If my use of a computer depended on biological evolution, and if I had invented the computer in, say, the mid-1980s, then at most one or two of my daughters (and their toddler offspring, my grandchildren) would currently be using computers. Instead, literally billions of people are doing so; not only that, but they are using technology that has been repeatedly updated, with new "generations" of hard- and software emerging every few years. During that time, of course, there has been effectively no change whatever in the biological nature of *Homo sapiens*.

In 1996, a nearly complete human skeleton—dated at between 8,400 and 8,690 years old—was discovered in Kennewick, Washington. Designated Kennewick Man, this individual became the focus of a struggle between Native Americans, who called him the Ancient One and wanted him buried with tribal honors, and paleoanthropologists, who wanted to study his remains. Twenty years later, the tribes won. For our purposes, a key point is that Kennewick Man wasn't really all that ancient; indeed, he is unquestionably modern in terms of his anatomy (and almost certainly, his physiology as well). Five years earlier, a mummified corpse had been discovered in a glacier in the Ötztal Alps along the border between Austria and Italy—hence the nickname "Ötzi," or "the Iceman." Radiocarbon dating revealed that Ötzi had lived between 3350 and 3100 BC, more than five thousand years ago—making him ancient indeed, but a newcomer compared to his Kennewick cousin.

For our purposes, the important thing is that neither Kennewick Man nor Ötzi is biologically different from modern human beings, but at the same time, there is no doubt that both of them were worlds apart from us, culturally. Imagine the following thought experiment: transfer one of Kennewick Man's or Ötzi's infant children to twenty-first-century Seattle or Bolzano, Italy, where Ötzi's mummified body now resides. These youngsters would undoubtedly grow up to be normal, unremarkable members of the current generation. Ditto for infants magically abducted from today's Internet culture in the Pacific Northwest or the ski tourism culture in today's Italian Alps and time-transferred to the surroundings in which Kennewick Man or Ötzi lived and to which they were doubtless well adjusted. By contrast, we can barely imagine the disaster if a comparable transfer were made between adults of these different times. Neither Ötzi nor Kennewick Man had ever experienced radio, television, computers, automobiles, firearms, flush toilets, books, and so forth: the list is nearly endless. And by the same token, it is unlikely—almost inconceivable—that an adult, no matter how much he or she flourished in the twenty-first century, would even survive in Neolithic times.

This discrepancy italicizes the vast gulf that has opened between our biological selves—changing very slowly, over great expanses of time—and our cultural selves, whereby for the last few thousand years we have been experiencing dramatic changes. Moreover, the rate of these changes has itself been changing—accelerating—in recent decades, even years and months. Recalling Kenneth Grahame's 1908 classic *The Wind in the Willows*, we have all been on Mr. Toad's Wild Ride. Exciting, yes. Downright exhilarating. But also frightening and dangerous. More manageable, perhaps, is to think of our situation as represented by Aesop's classic race between the tortoise (our biology) and the hare (our culture). In the myth, the tortoise wins, because the hare is overconfident, lazy, and downright foolish. In our human reality, neither can win because both exist within ourselves. Much of our success as a species has been due to the benefits derived by this bifurcated concatenation of tortoise and hare, biology and culture. But the paradox also pertains: even as we tortoises have ridden our hare-brained selves to great success, we are also threatened by these same successes, and especially by the fact that our tortoise nature has not evolved to deal effectively with the opportunities and risks presented by the hare.

HERE ARE A few examples, from the personal to the planetary.

Think about our species-wide sweet tooth. There are some people, of course, who don't like sugary substances, but they are distinctly in the minority. There is also little doubt as to the origin of this dietary preference. Primates all, we evolved in the trees, with a special predilection for fruits; not any old fruit, but those that are ripe, because ripeness indicates maximum nutritional value—a payoff provided by fruit-bearing trees, not out of the goodness of their woody old hearts, but as a lure to get arboreal creatures (including birds), to consume them, after which the seeds are pooped out elsewhere, thereby providing a selective payoff to trees that make rich, juicy, delectable fruits. And even if those fruits don't seduce fruit-eating foragers to spread the seeds within and the fruit falls uneaten to the ground, its high sugar content provides a nutritional boost to the resulting seedlings.

As a result, our ancestors engaged in a classic win-win biological exchange: we benefited fruit-bearing plants by spreading their seeds, and the plants benefited our primate forebears by making fruit consumption nutritionally worthwhile, with sugars being a benevolent lure. So far, so biological. Not surprisingly, ripe fruits and fruity flavors are still popular among modern *Homo sapiens*. Also to no one's surprise, people have seized on this sugar fondness via cultural practices that empower us to generate "sweets" that are nothing but sweet—that is, devoid of nutritional value, but superficially satisfying, and remunerative to chocolatiers, candymakers, bakers, the confectionary industry as a whole, growers and processors of sugar and high fructose corn syrup, and dentists. This saga presents a microcosm of the interaction of biological and cultural evolution, emphasizing how the rapid elaboration of the latter has taken advantage of the former, with the culinary-industrial hare presenting our Stone Age tortoise with temptations to which the latter is vulnerable.

A similar dietary situation obtains with regard to fats. Wild game is notoriously lean, and yet fats are calorically dense as well as having been relatively rare during 99 percent of our evolutionary past. When available, a fatty meal would have been highly valued. Today, not only can we pander to our fondness for "empty calories" in the form of highly sugared drinks and deserts but also we have the luxury of treating ourselves to highly marbled (and for many people, highly desired) steaks, courtesy of feed lots and industrial agriculture, along with the gustatorially appealing fat present in bacon, burgers, butter, cream, fried foods of all sorts, and so on. The beneficiaries here, in addition to the denizens of corporate animal husbandry, include—but aren't limited to—cardiologists, coronary bypass surgeons, diet consultants, coroners, and undertakers. And the culprits? Once again, the confluence of an ancient, historically and biologically adaptive behavioral predisposition (fondness for dietary fats) and a much more recent, culturally generated capacity whereby we can gratify that predisposition—to our great detriment.

A final additional factor—it is hard not to call it yet another nail in the coffin—concerns exercise. There are, of course, people who enjoy exercise, who go out of their way to get it, often paying large fees to join athletic clubs, gyms, hire fitness gurus, and so forth. But it is also notable how many gym memberships go unused, how many home exercising devices end up collecting dust in the attic or basement, and how serious is the increasingly worldwide obesity epidemic. This global health disaster is powered in large part by the fondness for sweets and fats just described, and further facilitated by humanity's widespread disinclination to get regular, vigorous exercise, despite the near-universal medical assessment that aerobic activity is beneficial for maintaining cardiovascular health, appropriate body weight, and protecting against a range of degenerative and inflammatory diseases.

Why don't people get more exercise? Once again, much of the blame resides in the disconnect between our biological past (and its imprint on our present-day predilections) and the circumstances and opportunities made available by our cultural interventions. There is debate among paleoanthropologists over precisely how much exercise our savannah-dwelling ancestors normally experienced, but it is unlikely that they took elevators or escalators; commuted by bus, train, or automobile; and spent their days at a desk or assembly line and their weekends as couch potatoes watching football games. For most of our evolutionary past, it was probably difficult to avoid getting heavy doses of exercise, and a good idea to minimize it whenever possible.

Our biology-culture disconnect is not limited to matters of nutrition, exercise, and health. Its imprint can be seen when it comes to nearly every big picture problem currently faced by our species. Take environmental destruction, including but not limited to greenhouse gas emissions and anthropogenic climate change. "Mankind has gone very far into an artificial world of his own creation," wrote Rachel Carson in *The Edge of the Sea*.

> He has sought to insulate himself with steel and concrete from the realities of earth and water. Perhaps he is intoxicated with his own power, as he goes farther and farther into the experiments for the destruction of himself and his world. For this unhappy trend there is no single remedy—no panacea. But I believe that the more clearly we can focus our attention on the wonders and realities of the universe about us, the less taste we shall have for destruction.[2]

At issue may be less a "taste for destruction" than a biologically instilled insensitivity to it, along with an equally biological inability to control our own destructiveness. By contrast, animal and plant species lack the opportunity to literally destroy whole ecosystems; hence, they have not been outfitted by evolution with restraints against doing so. This applies, as well, to a certain weak-bodied primate who used to wander around the African continent. But cultural evolution rapidly outpaced biology as human beings developed increasingly complex and environmentally destructive technologies, not only creating and spewing toxic substances but also clear-cutting forests, destroying soils, emptying aquifers, exterminating other species, and so forth. Insofar as natural selection might operate at the level of planets, such behavior would eventually result in the Earth itself being rendered uninhabitable by its human occupants, ultimately—if this were possible—to be replaced by other planets whose inhabitants behaved more in synch with their own long-term best interests.

These examples don't nearly exhaust the ongoing conflicts between human culture and biology, and the resulting extent to which our cultural power has been obtained at the cost of paradoxical threats to our own well-being. Overpopulation, which lies at the root of many problems, is mostly due to the fact that human reproductive biology (like that of all other organisms) is set at its maximum effective rate—which through most of human prehistory didn't result in excessive numbers because death rates were also high. But within historic times, culturally mediated intervention has substantially reduced death rates—through effective public health programs, improved food-rearing practices, and so forth—while the biological inclination for maximizing reproduction has not taken this fundamental modification into account. (Fortunately, cultural adaptations exist to remedy this situation, namely, birth control; unfortunately, however, such interventions are typically opposed by other cultural phenomena, namely fundamentalist religions and their political enablers.)

Things are more immediately dire yet, with respect to violence and weapons. We have already noted that although war is not in our genes, aggressiveness is. More accurately, we are capable of interpersonal aggression and even violence, under certain conditions. Unlike many other animal species, however, we are not efficient killing machines, based simply on our biological endowment. We don't have sharp horns or antlers, lethal canines, poison-injecting fangs, claws designed to rip and tear. It is very difficult, in fact, for a naked, untrained human being to kill another member of our species.

Konrad Lorenz made much of the observation that most animals that naturally possess lethal equipment also possess—just as naturally—inhibitions against using this equipment against conspecifics. We now realize that Lorenz's interpretation was somewhat exaggerated.[a] Such inhibitions are not absolute: wolves, for example, do occasionally kill other wolves. Ditto for lions and other animals possessing lethal armamentarium. But the generalization is nevertheless valid, in that it is generally true: by and large, lethally armed creatures are also strongly disinclined to kill each other. Where does that put human beings? We have already noted that our species, by contrast to "natural born killers" of the animal world, is poorly equipped to kill other humans, based on our biology alone. But once again, unlike other species, and thanks to our cultural and technological innovations, our lives are no longer circumscribed by biology alone—a biological-cultural mismatch that is especially pronounced, and particularly tragic, when it comes to weaponry.

For thousands of generations, even after our ancestors developed the ability to hunt, maim, and kill with devices beyond their unaided bodies, they were limited to

---

[a] As is, incidentally, his kindly reputation as avuncular old Konrad, followed by a waddling gaggle of imprinted geese; the sad reality is that Lorenz, an Austrian, promoted Nazi ideology and atrocities during World War II. Although it is fair to hold this despicable history against his reputation, it needn't justify ignoring Lorenz's legitimate scientific achievements as one of the founders of ethology, the biological study of animal behavior.

muscle-powered weapons, beginning with sticks and stones, "progressing" from cudgels to edged tools (crude knives, swords, darts, spears, bows and arrows), with an accelerating array of increasingly lethal killing implements, proceeding from blunderbusses to flintlock and matchlock rifles, cannons, machine guns, explosive artillery, submarines, airplanes, napalm and chemical weapons, rockets and missiles, and—the ultimate in lethality—nuclear bombs and warheads.

This transition has been breathtaking in many ways: in the speed and efficiency with which killing can now be accomplished, and in the number of victims encompassed, along with the ease of carrying out the action. Compare, for example, the physical effort required to kill someone with your bare hands, to using a club, to a knife or spear, to squeezing the trigger on a pistol or rifle, to a bombardier operating a lever, to a launch control officer sitting in an ICBM silo who needs only to turn certain keys and switches in a predetermined sequence to obliterate most of a continent.

Crucially important for our purposes—and perfectly consistent with the hare-and-tortoise argument presented earlier—is the extraordinary speed with which these various innovations in weaponry have been developed, all by a process of cultural evolution that has been blindingly fast compared to its biological counterpart. Theodor Adorno once quipped, "No universal history leads from savagery to humanitarianism, but there is one leading from the slingshot to the atom bomb."[3] This history wasn't actually universal, in that it occurred in different societies at different times. Nonetheless, it was overwhelmingly cultural, and it proceeded at a rate that outpaced our much more universal biological evolution, none of which prepared our species for how to cope with nuclear weapons.

As rattlesnakes evolved hypodermic fangs and lethal venom, they experienced a history that was indeed universal for all members of their species, and that occurred slowly, over evolutionary time, during which they also evolved substantial inhibitions regarding the use of their lethal armament. When two rattlesnakes fight, they do not seek to bite each other; rather, each struggles to push the other over backward, after which the loser slithers away—to fight, nonlethally, another day. But human beings never acquired lethal weapons characteristic of rattlesnakes in a comparable biological manner. As a result, we are, in a sense, less dangerous than rattlesnakes—based on our anatomy alone. But cultural evolution has endowed our species with weapons that are immensely more threatening than rattlesnake fangs. And making matters immensely worse, because our weapons were acquired overnight (as Darwinian evolution reckons these things), there has been insufficient time for our slow-moving biology to endow us with behavioral inhibitions that would be appropriate to the weapons that we have acquired through cultural evolution. Lacking the fangs of a rattlesnake or the claws of a tiger, we also lack their inhibitions about using these things. We are, via our cultural evolution, in over our biological heads.

In no other dimension is there a more dramatic or more terrifying paradox of culturally mediated power wielded by a creature that is not only biologically unprepared to do so, but actively ill-prepared. This is manifested in many ways. For example, our biological psyche deals relatively well with the reality of individual death; our ancestors encountered mortality often and pretty much exclusively on a personal scale. In this respect, numbers are numbing. Our minds and our emotions boggle and rebel at the death of millions. What we cannot comprehend we tend to ignore or deny, or at minimum, seek refuge in more manageable phenomena. Similarly with the actual effects of nuclear detonations: to our primitive, biologically honed primate brain, "hot" conjures up a blistering summer day, or perhaps boiling water or a campfire, nothing even remotely approaching the millions of degrees (exceeding the temperature at the surface of the sun) produced by a single nuclear explosion. Unable to mentally encompass this reality, we resist taking it seriously.

By a similar kind of bio-logic, we respond meaningfully to immediate threats such as a murderous individual, a thunderstorm, or a nearby fire, but shut down awareness of threats that are genuinely real, but seem unreal to our myopic hominid mind. Moreover, since nuclear weapons cannot readily be tasted, seen, felt, or heard, they tend to lack psychological reality, compared to, say, a snarling dog or a drunken driver. And the fact that the US government, for example, consistently refuses to "confirm or deny" the presence of these weapons, anywhere, only adds to the temptation for such denial. This is only a partial list. By now, the general problem should be clear: not only has our cultural evolution provided us with implements of terrifying destructive power, but the situation is all the more terrifying because in the process, cultural evolution has confronted human beings with a disinclination—in many cases, an actual incompetence—to perceive these weapons accurately, and thus, an inability to deal with them as they deserve.

Here is a suitable metaphor, seemingly far removed from nuclear weapons. Among the more notable inhabitants of the northern arctic are musk oxen, great shaggy hoofed mammals whose *qiviut* fur is extremely soft, warm, and valuable. Their meat is also reputed to be tasty, wherein lies part of their problem. For tens of thousands of generations, musk oxen have had to confront fearsome predators, notably wolves. They evolved an effective response, forming a pattern like the spokes of a wheel, with the adults facing outward and the vulnerable juveniles inside. With such a configuration, no matter how they approach, wolves faced a formidable array of sharp and sturdy horns, backed up by a half-ton of raw and angry pot-roast. For eons, this counter-strategy has served musk ox needs. But in modern times, the situation has changed. Today, the primary threats to musk oxen are no longer wolves, but human beings, riding snowmobiles and armed with high-powered hunting rifles. Musk oxen, however, continue to respond as they always did in the past, forming their trusted, immobile defensive pattern, a behavior that worked well against wolves but which is currently not only ineffective, but probably the worst thing they could do. (The best would likely be to scatter and run as fast as they can.)

In August of 1946, a year after the atomic bombing of Hiroshima and Nagasaki, Albert Einstein sent a telegram to the world's most renowned physicists, urging them to organize in opposition to nuclear weapons. "The splitting of the atom," he wrote, "has changed everything but our way of thinking, and thus we drift toward unparalleled catastrophe." We are worrisomely like musk oxen, for whom the invention of snowmobiles and hunting rifles changed everything but their way of thinking, leaving them at risk of extinction. Thanks—no thanks!—to our invention of nuclear weapons, we have presented ourselves with hunting rifles writ large, to which we insist on responding like stubborn musk oxen, dealing with this new and fearsome culturally evolved danger as though it simply represents another traditional threat, and responding with variants on our old tried-and-true biologically evolved responses. Instead of arranging ourselves like musk oxen, most people refrain from thinking about nuclear weapons, and thus hardly respond at all. At the same time, those who focus professionally on nuclear bombs and warheads tend to think of them as just "normal" weapons, such that having more of them is better and will somehow make us safer, especially if this means more than our opponents. Our stubborn internal musk ox (or biologically evolved inner tortoise) is largely unwilling or unable to wrestle creatively with the products of our fast-moving technological hare. Hare-brained indeed.

THE EVOLUTIONARY BIOLOGIST Julian Huxley was part of an eminent English family, whose most notable patriarch was the great nineteenth-century anatomist and social philosopher Thomas Henry Huxley (aka "Darwin's bulldog"). One of T. H. Huxley's grandsons was the writer Aldous (author of *Brave New World*, among other masterpieces); another, Andrew, was a brilliant neurophysiologist whose research earned him a Nobel Prize. Yet another grandson was Julian Huxley, an eminent scientist in his own right, the first director of UNESCO, and the title of whose magisterial book *Evolution the Modern Synthesis* was the model for E. O. Wilson's equally consequential book *Sociobiology: The New Synthesis*. At one point, Julian warned against "nothing butism," the tempting conclusion that just because human beings are animals (which Julian didn't dispute for a moment), we should not conclude that we are *nothing but* animals.

Among the main themes of the present book, and especially Part II, is that we are indeed animals. On the other hand, a major theme of social science has long been that we are special, not so much by virtue of cosmic centrality or because we are the apple of god's eye, but because of the extraordinary power of human culture, which has ostensibly lifted us into a new realm of being human. But has it really? At the conclusion of Arundhati Roy's stunning novel *The Ministry of Utmost Happiness*, a transgender woman has taken a toddler for a walk in night-time Delhi. The little girl has to urinate, after which she "lifted her bottom to marvel at the night sky and the stars and the one-thousand-year-old city reflected in the puddle she had made." It is a remarkable image, worth meditating on: an immense human-made city, ancient yet vibrant, iconic example of cultural artifice, reflected in a deeply biological puddle of a child's pee. The profound impact of human culture—more usefully seen as cultural evolution—is undeniable. Whether it has ushered us into an entirely new realm, however, is debatable. More likely, it has provided us with immense opportunities, but also immense dangers and difficulties, especially given that its speed and innovation has been paradoxically paired with the painfully laggard pace of biological evolution.

In the Monty Python movie *The Life of Brian*, one of the characters—trying to evoke a rebellion against the Roman Empire—asks, "Apart from the sanitation, the medicine, education, wine, public order, irrigation, roads, the fresh-water system, and public health, what have the Romans ever done for us?" (He wasn't successful.) Substitute culture for the Romans, and few people would willingly give up the benefits that cultural evolution has wrought and turn our backs on antibiotics, life-saving surgery and public health measures, convenient communication and transportation, our pets, our smartphones, or our computers. At the same time, it is downright irresponsible to deny the costs that have accompanied our bifurcated existence as both primitively biological and powerfully cultural creatures.

Which leads to this iconic observation to young Peter Parker from his wise Uncle Ben, advice that is not limited to Spiderman: "With great power comes great responsibility."

Old paradigm: By virtue of our uniquely human cultural and technological achievements, we have raised ourselves above mere animals and even above natural limitations.

New paradigm: We are the products of biological and cultural evolution, a combination that has endowed us with extraordinary power; at the same time, however, these two processes are often out of synch, a disparity that confronts us with extraordinary difficulties as well as challenges.

NOTES

1. M. McCarthy, *The Moth Snowstorm: Nature and Joy* (New York: New York Review of Books, 2016).
2. Rachel Carson, *The Edge of the Sea* (Boston: Houghton Mifflin, 1955).
3. Theodor Adorno, *Negative Dialectics*, 2nd ed. (New York: Bloomsbury Academic, 1979).

Conclusion
Optāre aude

DEPRIVED OF SO many of our prior people-based paradigms, many of them comforting, what's left? If we aren't the center of the cosmos, weren't specially created by a benevolent god, aren't even separate and distinct selves, don't have a monopoly on consciousness and rationality, lack free will, aren't in charge of our own actions, if we behave "altruistically" because of biological factors rather than divine grace, if we can't lay claim to being naturally monogamous, or peaceful (or war-disposed, either—something of a relief!), can't rely on being the specially created apples of god's eye, nor on reliably telling the truth, or on being especially well constructed or even guaranteed to be benevolently disposed toward our own children, and no longer able to clothe ourselves in the illusion that our lives necessarily have meaning, where does that leave us?

Sometimes it is said that the truth hurts. And sometimes it does. Far more likely, however, and more importantly, the truth *can* hurt—but it can also be healing and is for the most part downright necessary. This book has sought to identify an array of truths about old and new paradigms concerning the human place in the cosmos and about human nature itself. Sometimes it has hurt; I would like to think that more often it offers the prospect of liberation. And if nothing else, it has the virtue of being, well, true.

This is being written in mid-2017, when the president of the United States has been persistently and brazenly revealing not only an indifference to the truth but also an actual predisposition to promote outright lies. Moreover, when called on these lies, he hasn't hesitated to double down on his own blatant falsehoods. It isn't clear whether he is intentionally promoting what he knows to be untrue or is genuinely deluded and thus unable to distinguish between objective truth and "alternative facts" that he mistakenly believes correspond to reality. In any event, insofar as he succeeds in getting away with these and other outrages on facticity, he threatens to substitute a new paradigm consisting of a persistent tissue of lies, in place of an older one, in which what we might unblushingly call the truth has long had pride of place.

Idealistic and downright corny as it might seem, there is much to be said for the truth, the whole truth, and nothing but the truth—even if on occasion it hurts. At one level, admittedly a relatively minor one, there is the problem of punctured dreams and expectations. "Often I have thought of the day when I gazed for the first time at the sea," wrote Thomas Mann. "The sea is vast, the sea is wide, my eyes roved far and wide and longed to be free. But there was the horizon. Why a horizon, when I wanted the infinite from life?"[1] Wanting the infinite is a bit much. We all need to make accommodations with the truthful horizons that circumscribe each of our lives. We will never know what it would have been like to talk with Isaac Newton, to bestride the world like a colossus, or to experience life as an octopus or a bat, painful as that realization might be (at least for some of us).

Seemingly well informed people blather about somehow transcending our limitations, by "evolving" into new, posthuman, part-cyborgs, or becoming an "interplanetary species" that colonizes other worlds as a way of escaping resource depletion, overpopulation, ecological collapse, or possible viral pandemics—not to mention nuclear war. Kenneth Patchen's poem "Credit to Paradise" speaks especially to such fantasies, via these lines:

> Isn't all our dread a dread of being
> Just here? Of being only this?
> Of having no other thing to become?
> Of having nowhere to go really
> But where we are?[2]

The alternative to boundless, horizon-free knowledge or dread of being just here, of being only this, and of having no other thing to become or place to go isn't ignorance but a potentially disastrous injection of social and political propaganda where science and respect for facts should hold pride of place. In George Orwell's novel *1984*—which, not coincidentally, experienced a surge in readership following Donald Trump's election in 2016—the hero, Winston Smith, vows to adhere to truth as he sees it. He avers, "The solid world exists, its laws do not change. Stones are hard, water is wet, objects unsupported fall toward the earth's center." Not all old paradigms are invalid; some old saws retain their sharp teeth. The one that argues for the existence of certain verifiable and empirical realities—the stuff of science—is not only true, but immensely valuable.

The laws of physics—manifest, say, in the case of global warming or nuclear winter—will always take precedence over the vagaries of politics.

There can nevertheless be a certain grandeur in illusions, especially those that delude ourselves into thinking that we are better, nobler, worthier than we are—particularly if those "positive illusions" lead to a self-fulfilling prophecy that leaves us being in fact better, nobler, worthier than we might otherwise be. Don Quixote is a useful model in this regard: mad, disconnected from reality, and thus ridiculous, the Don nonetheless achieves a kind of nobility by virtue of his disconnection from an otherwise grubby reality. In Part 2 of Cervantes's novel, Quixote is ultimately overthrown by one Sansón Carrasco, masquerading as the "Knight of the Mirrors," who challenges the Don to a duel and is unhorsed and almost killed. Later, he again meets Carrasco, now masquerading as the "Knight of the White Moon," who this time defeats the old man and forces him in penance to return home and give up his fantasies of knight-errantry for one year. Quixote does so, and, regaining his senses, goes back to being the pedestrian but "normal" Alonso Quihana, who then dies, rather abruptly—a disappointing anticlimax to a great work of fiction, widely considered the first modern novel.

This will doubtless outrage literary traditionalists, but a more intriguing as well as satisfying conclusion is provided by the 1964 Broadway musical, *Man of La Mancha*. Here, our quixotic hero encounters the "Knight of the Mirrors" (Sansón Carrasco, again) just once and is defeated, largely because those mirrors force Don Quixote to see himself as he really is, a ridiculous, elderly lunatic. It isn't appealing to be cast as a Carrasco—a self-serving physician in love with Alonso Quihana's niece, both of whom find the escapades of Don Quixote to be socially demeaning—and who emerges as the musical's villain because he forces our hero to rub his nose in the painful reality of his own pedestrian life, devoid of romantic pretensions of self-importance.

But a windmill is not a four-armed giant, the prostitute and scullery maid Aldonza is not the beautiful and chaste Dulcinea, a shaving bowl is not the Golden Helmet of Mambrino—even if it requires a cynical Knight of the Mirrors to convince us of these facts. The truth, albeit painful, has a habit of emerging and—yes—even offers the possibility of setting us free.

Denied the comforting crutch of illusions, we have the opportunity and obligation to do something truly extraordinary: to see ourselves as we really are, and then, to use the insights to understand ourselves better, and maybe even to behave not only with humility but even more nobly and humanely. "The greatest mystery," wrote Andre Malraux, "is not that we have been flung between the profusion of the earth and the galaxy of the stars, but that in this prison we can fashion images of ourselves sufficiently powerful to deny our nothingness."[3]

Even though we aren't the center of the universe—or even of the solar system—and even though we aren't nearly as special as many might like, we can take comfort in Alfred Tennyson's suggestion in his poem, "Ulysses," that

> though much is taken, much abides, and though
> We are not now that strength which in old days
> Moved earth and heaven, that which we are, we are.

Moreover, "that which we are" is impressive enough, without any illusions of extra-specialness. Choosing to know ourselves, we then have at least the possibility of choosing our future, mindfully and with maximum insight. As the Israeli American scholar Uriel Abulof suggests, we might want to consider adding to *Sapere aude* ("dare to know," which Immanuel Kant proposed as the motto for the Enlightenment) the existentialist variant *Optāre aude*: "dare to choose."

Why is daring required? Because, as Professor Abulof points out, it can be perilous to seek and find something vital and important, and the more vital and important, the more perilous is its pursuit:

> In the Arthurian Legends, Merlin the magician engraved onto the chairs of the Round Table, in gold leaf lettering, the name of each of the Knights, save one chair on which he inscribed *Siege Perilous*, the "Perilous Seat." This chair was reserved for the Knight who was destined to seek and find the Holy Grail. For anyone else the Perilous Seat was fatal. *Siege Perilous* is the state of people—and peoples—in a secular age.[4]

It goes almost without saying that a secular age is also a scientific one, in which the Holy Grail is no longer a mythical sacred object, but the overthrow of misleading paradigms and their replacement by brighter, clearer, more accurate ones.

At the end of Ray Bradbury's *Martian Chronicles*, a human family—having left Earth to avoid impending nuclear war—looks into one of the "canals" of their new home, hoping to see Martians. They do: their own reflections. Such images—especially when provided by the best of modern science—are a far cry from Malraux's "nothingness." Moreover, the loss of illusions is testimony to the vibrancy of science itself, and to the regular, unstoppable enhancement of rational understanding as we approach an increasingly solid grasp of how the world is put together, and what it means to be human.

NOTES

1. Thomas Mann, *Little Herr Friedemann* (Berlin: S. Fischer Verlag, 1898).
2. Kenneth Patchen, "Credit to Paradise," in *Selected Poems* (New York: New Directions, 1957).
3. Andre Malraux, *The Walnut Trees of Altenburg* (Chicago: University of Chicago Press, 1992).
4. Uriel Abulof, *Abyss and Horizon* (forthcoming, Cambridge University Press).

Index

Numbers followed by n indicate notes.

absurdity, 27, 29–31
Abulof, Uriel, 193
action at a distance, 90
Adams, Douglas, 25, 39
Adorno, Theodor, 186
Age of reason, 104
aggression, 142, 146–161, 185
alchemy, 10
Aldrich, Henry, 104
aether, 5
Alex (African gray parrot), 93, 100
algometric analysis, 13
alternative facts, 190
altruism, 12, 116, 125, 154, 162–171,
    176–178, 190
amenorrhea, lactational, 122
anaerobes, 58
anatman (non-self), 68, 75
angels, 162–171
animals. *See also specific animals*
    behavior, 101, 139, 162
    cognition, 98, 101–102
    communication, 125–134
    intelligence, 93, 95–109
    parent–offspring conflict, 111–115
anthropic principle, 39–51
Anthropocene era, 17–18, 20
anthropocentrism, 18–19, 86
anthropodenial, 18
anthropodiminution, 6, 18, 179
anthropology, 84, 146–148, 150
    cultural, 14, 84
    evolutionary, 98–99
    paleoanthropology, 151, 184
antihero, 104–105
apes, 136
Ardrey, Robert, 147

Aristotle, 7, 28, 54–55, 103
Arnol'd, Vladimir, 54
art, 87, 89
Asimov, Isaac, 7
asphalt, 55–56
astronomy, 22, 42
astrophysics, 45
atheism, 5
atom bomb, 109, 186, 188
ATP (adenosine triphosphate), 78
authenticity, 12
autonomy, 172–178

Bach, Johann Sebastian, 87
Bacon, Francis, 19
bacteria, 55–61
Batek people, 149–150
Beckett, Samuel, 29–30
bees, 13–14, 126–127, 156
behavior
    altruistic, 12, 116, 125, 154, 162–171,
        176–178, 190
    animal, 101, 139, 162
    communicative, 125–134
    human, 14, 85, 105
    moral, 167–170
    risk averse, 106–107
    sexual, 121–122
    spiteful, 105
    symbol, 14
    warring, 146–161
behavioral dimorphism, 136
Bekoff, Marc, 98
Belloc, Hilaire, 94
beneficence, 27, 162–171
Benjamin, Walter, 91
Bergson, Henri, 57

Bernier, Abbé, 86
Berzelius, Jacob, 57
Bible, 8, 23, 36, 86, 94
Bierce, Ambrose, 44
Big Bang theory, 10
Big Crunch, 41
biological evolution, 179–189
biology, 33–34, 77–78, 90, 177–178
 evolutionary, 28, 31, 72–73, 85, 112,
  123–126, 150
 human, 9, 77–78, 138, 148, 177–189
 invertebrate, 173
 microbiology, 10–11
 mismatch with culture, 179–189
 neurobiology, 77, 102
 sociobiology, 148
 Soviet, 64
bipedalism, 36n
black holes, 48
bloodletting, 10
Bloom, Paul, 100
Boehm, Christopher, 151
Bohr, Niels, 88
Bolsheviks, 64
Bonobos, 151
border collies, 99–100
botany, 102
Bowlby, John, 111
Boyle, Robert, 88
Bradbury, Ray, 193
Brahe, Tycho, 3, 14, 47
brain size, 12–13
brain worms, 174
Brecht, Bertholt, 168
Bridgewater Treatises (Royal Society of
 London), 33
Brobdingnagians, 16–17
Brooke, Rupert, 40
Bruno, Giordano, 23
Buchsbaum, Ralph, 173
Buddhism, 5, 23–24, 69–70, 75
Bulgakov, M., 64
bull elephants, 135–136

Calvin, John, 146
camouflage, 126
Camus, Albert, 30–31
Cannon, William, 53–54

Carson, Rachel, 184
Carter, Brandon, 40–41
Carter, Jimmy, 130
Cartmill, Matt, 12–13
cartoons, 111
Catholic Church, 23
causation, 28
centaurs, 65
centrality, human, 1–81
Cervantes, Miguel de, 192
Chagnon, Napoleon, 147–149
Cherokee, 160
Chicxulub asteroid, 46
children
 parent–offspring conflict, 111–124
 sibling rivalry, 114–116, 123
chimeras, 63–67
chimpanzees, 98–99, 101, 152
chimphumans, 63–67
choice, 190–193
cholera, 175
Christianity, 23, 86, 168
Cleopatra, 105
climate change, 2, 17–18, 179
cloning, 11, 63
cognition, complex, 93–110
cognitive ethology, 93, 97–98
Coleridge, Samuel Taylor, 91–92
Collins, Francis, 162, 170
Columbus, Christopher, 88
communication, 125–134. *See also* language
Communist Manifesto (Marx and Engels), 21
completely ridiculous anthropic principle
 (CRAP), 43
complex thought, 93–110
conflict
 parent–offspring, 111–124
 sibling rivalry, 114–116, 123
 weaning, 114–115
continental drift, 20
Coolidge effect, 138
Copernican Mediocrity, 50
Copernicus, Nicolaus, 3, 5, 22, 28,
 40–41, 47, 88
copulation, extra-pair, 130
copulatory interference, 122
Correns, Carl, 88
cosmic perspective, 79

Cosmides, Leda, 109
cosmology, 46–47
Cosmos (Sagan), 91
cousins, 117–118
Crane, Stephen, 105–106
CRAP (completely ridiculous anthropic
        principle), 43
creationism, 95–96
creative arts, 87
Crick, Francis, 172
CRISPR, 63, 65
Critchley, Simon, 147
Crutzen, Paul, 17–18
cry characteristics, 119–120
cryophiles, 58
cultural anthropology, 14, 84
cultural mutations, 181
culture, human, 179–189
cybernetics, 73

Dalai Lama (Tenzin Gyatso), 75
Daniel, 104
Dart, Raymond, 146
Darwin, Charles, 4, 13–14, 20–21, 31, 34–35,
        88, 162–163, 167–168
Darwinian Mediocrity, 50
da Vinci, Leonardo, 87
Dawkins, Richard, 5, 26, 50, 55, 66n, 73, 127,
        164, 176
death, 27
deception, 125–134, 139, 190
de la Bruyere Jean, 168
Dennett, Daniel, 5
Descartes, René, 44
design, intelligent, 33–38
Deutsch, David, 89–90
developmental psychology, 122
DeVries, Hugo, 88
de Waal, Frans, 98, 100–103
d'Holbach, Baron, 86
Dicke, Robert, 41
Dicrocoelium dentriticum, 174
dimorphism, 136
        behavioral, 136
        sexual, 135–136
dinosaurs, 46, 79
Diplozöon paradoxicum, 141
discovery, 88

disenchantment, 3
dishonesty, 125–134
Disney, Walt, 111
DNA, 78
DNA repair, 60
DNA transfer, 60
Dobzhansky, Theodosius, 34
Doctorow, E. L., 132
Dolly the sheep, 11
Donne, John, 22, 79
Don Quixote (Cervantes), 192
Dostoyevsky, Fyodor, 104–105
double slit experiment, 48

eavesdropping, 132
Ecclesiastes, 94
ecology, 54, 69–70
ecosystems, 70–72
efficient causes, 28
Egerton, Francis Henry, 33
Einstein, Albert, 5, 41, 43–44, 79, 88, 188
élan vital, 57
electromagnetism, 10
Eliot, T. S., 15
embryology, 11
Emerson, Ralph Waldo, 61, 69
endosymbiosis theory, 78
Engels, Friedrich, 21
Enlightenment, 7, 22, 104
Enquiry Concerning the Human (Hume), 85
environmental science, 17
ethics, 54–55, 163
ethology, 93, 97–98, 126, 185n
eukaryotes, 78
evolution, 4, 8, 13–14, 20–21, 25–26, 34–36
        biological, 179–189
        cultural, 179–189
        human, 86, 107, 179–189, 191
evolutionary anthropology, 98–99
evolutionary biology, 28, 31, 72–73, 85, 112,
        123–126, 150
evolutionary cognition, 97–98
evolutionary existentialism, 27
evolutionary fitness, 148
evolutionary psychology, 109, 148
examined life, 8
exercise, 183–184
existential freedom, 12

existentialism, 27, 31
The Extended Phenotype (Dawkins), 73
extra-pair copulation, 130
extraterrestrial life, 23–24, 48
extremophiles, 55–61

facticity, 190
fallibilism, 89–90
fathers, 121–123
fats, 183
favoritism, 164
Federalist Papers, 153
female orgasm, 140
Feynman, Richard, 48
The Fierce People (Chagnon), 147–148
final causes, 28
fitness, 111, 113, 162–163
    evolutionary, 148
    inclusive, 163–164
    indirect, 164
    reproductive, 163–164
flatworms, 141
fragility, 55
Francis, (Pope), 86
freedom, existential, 12
free parameters, 43–44
free will, 172–178
Freud, Sigmund, 4, 20, 89, 122–123, 140, 144
friendship, 166
Frost, Robert, 76, 91
fruit, 182–183
Fry, Douglas, 150
Fukuyama, Francis, 89
fundamentalism, 86
Futuyma, Douglas, 28

Galileo Galilei, 3–5, 7, 22, 47
Ganesh, 65
Garcia, Hector, 142–143
Gardner, Martin, 43
genes and genetics, 26, 66, 72–74, 88,
    139–140, 162, 176–178
    natural selection among, 153
Genesis, 36
geology, 28
giraffes, 36
"God of the gaps" argument, 49–50, 169–170
god talk, 43–44

Goethe, Johann Wolfgang von, 76–77
golden interval, 41
Goldilocks Principle, 41
Goodall, Jane, 120, 122
Gorky, Maxim, 64
Goya, Francisco, 109
Grahame, Kenneth, 96, 182
Gray, Asa, 34
Gray, John, 147
greenhouse gases, 17–18
Griffin, Donald, 98
group selection, 153–154, 167–170
Gulliver's Travels (Swift), 16–18, 103
guns, 94
Gyatsu, Tenzin (Dalai Lama), 75

Haldane, J. B. S., 60–61
halophiles, 58
Hamilton, William D., 163
handicap principle, 133
Hans, Clever (horse), 101
happiness, 103
Harris, Sam, 5
Hawking, Stephen, 41, 43
Heartburn, 141
"Heaven" (Brooke), 40
Heimlich maneuver, 38
Heine, Heinrich, 26
Heisenberg, Werner, 88
Helicobacter pylori, 11–12
Henley, William Ernest, 172
Heraclitus, 4–5, 19
heroism, 30–31, 167–168
Hertz, Heinrich, 10
Hinduism, 65
history, 89
Hitchens, Christopher, 5
Hitchhiker's Guide to the Galaxy
    (Adams), 25, 39
Hobbes, Thomas, 158
Holocaust, 109
homosexuality, 11, 143–144
honesty, 125–134
host manipulation, 27n, 173–175
Hoyle, Fred, 10
Hoyle state, 42–43
Hubble, Edwin, 41
Hugo Award, 59

humanitarianism, 186
humanzees, 63–67
Hume, David, 34, 77, 85, 104, 144
Hutton, James, 28
Huxley, Aldous, 146–147, 188
Huxley, Andrew, 188
Huxley, Julian, 57, 188–189
Huxley, Thomas, 112
Huxley, Thomas Henry, 188
Huygens, Christiaan, 88
hybrids, 65–66, 66n

the Iceman (Ötzi), 181–182
id, 144
illusion, 191–192
inclusive fitness, 163–164
indirect fitness, 164
individuality, 68–69, 81, 178
industrialization, 18
infantile or childhood sexuality, 121–122
infra-terrestrials, 56
inhibitions, 185
Inquisition, 23
In Search of Lost Time (Proust), 84
intellectual faculty, 13–14
intelligence, 93–109
intelligent design, 33–38
interbeing, 69–70
intergenerational conflict, 111–124
intergroup violence, 155
interpersonal morality, 166–168
interpersonal violence, 142, 146–161, 185
intuition, 7
invertebrate biology, 173
in vitro fertilization, 66
Islam, 23, 86
Ivanov, Ilya Ivanovich, 63–64

Jay, John, 153
jealousy, sexual, 140
Jefferson, Thomas, 19, 27–28
Jeffers, Robinson, 169
Johnson, Samuel, 86
Judaism, 53, 86

Kagan, Jerome, 111
Kahneman, Daniel, 106–107
Kant, Immanuel, 193

Kelvin, Lord, 10
Kennewick Man, 181–182
Kepler, Johannes, 3, 22, 47
Khayyam, Omar, 103
Kilmer, Joyce, 75
King, Martin Luther, Jr., 72
kin selection, 164, 167–170, 176–178
knowledge, 89–90
Koch, Robert, 10–11
Krebs, John, 127
Krutch, Joseph Wood, 92
Kuhn, Thomas, 5–6

labradoodles, 65
lactational amenorrhea, 122
language, 14, 149. *See also* communication
Laplace, Pierre, 170
Laputans, 103
laryngeal nerves, 37
learning, 13
Lehrer, Tom, 128
Leibniz, Gottfried, 45, 88
Lenin, Vladimir, 64, 175
Leonardo da Vinci, 87
life
    definition of, 54–55
    first appearance on Earth, 56
    interconnectedness of, 70–72
    meaning of, 25–32
    reverence for, 52–53
    right to, 67
    toughness of, 52–62
life cycles, 70–71
The Life of Brian, 189
Lilliputians, 16
Lincoln, Abraham, 162
Linnaeus, Carl, 94–95
Lister, Joseph, 10–11
literature of the absurd, 29
Liu Cixin, 59–60
Lorenz, Konrad, 87, 97, 126, 185, 185n
Los Angeles Review of Books, 147
Lowell, Percival, 10, 23
Lyell, Charles, 102

Mach, Ernst, 88
Machiavelli, Niccolò, 131
machine guns, 94

Malraux, Andre, 192
Mann, Thomas, 191
Man of La Mancha, 192
Mardu people, 150
Margulis, Lynn, 78
Marianas Trench, 55–61
Marshall, Barry, 11–12
Marxism, 21
Marx, Karl, 21
masochism, 105
Maxwell, James Clerk, 10, 88
Mayr, Ernst, 150
meaning, 193
    search for, 25–32
mechanics, 20
medical science, 10–12
mediocrity, 50
Meißner, Ulf-G, 42–43
Mencken, H. L., 159
mendacity, 130–133
Mendel, Gregor, 88
mental capacity, 93–110
methanotrophs, 58
microbiology, 10–11
"Mikado" (Gilbert and Sullivan), 34
Miller, Stanley, 56
Milton, John, 9
mind reading, (theory of mind), 131
*The Ministry of Utmost
        Happiness,* 189
Minnesota Fats, 34
misanthropic principle, 50
misanthropy, 146
Mishnah Sanhedrin, 53
mitochondrial DNA, 78
mitosis, 178
modocentrism, 19–20
Mohammed, 23
monogam-ish, 140–141
monogamy, 12, 135–145
monotheism, 142–143
monsters, 109
Montaigne, Michel de, 104
morality, 162–163, 167–170
Moral Law, 162, 170
Morgenbesser, Sidney, 45
Morley, John, 100
Mormons, 23

Morris, Desmond, 101
mothers, 121–123
multiverse hypothesis, 47
Mumford, Lewis, 3
Mundurucu people, 155
Murphy (Beckett), 29
musk oxen, 187–188
mutations, cultural, 181
"The Mysteries" (Goethe), 76–77

Napoleon Bonaparte, 170
Nash, Ogden, 120–121
National Aeronautics and Space
        Administration (NASA), 42
National Institutes of Health
        (NIH) (US), 66
The National Interest, 147
Native Americans, 181
natural selection, 20–21, 34, 72–73, 113,
        162–163, 169, 174
    cosmological, 48–49
    levels of, 153
nature
    human, 83–193
    human place in, 1–81
Nature (Emerson), 69
nematodes, 173
nepotism, 164, 176–178
neurobiology, 77, 102
neurosis, 144
new atheism, 5
Newman, Paul, 127
New Scientist, 147
new Soviet man, 64
Newton, Isaac, 21–22, 44, 88, 90, 104, 106
The New Yorker, 12
Nietzsche, Friedrich, 22–23, 76, 130, 168
Nixon, Richard, 130
Nobel Prize, 11–12
non-self *(anatman),* 75–76
nothing butism, 188
Nowak, Martin, 169
nuclear DNA, 78
nuclear physics, 42–43
nuclear weapons, 109, 186, 188

observational learning, 13
Occam's Razor, 97–98

Oedipal conflict, 112
Oedipus complex, 4
offspring
    parent–offspring conflict, 111–124
    sibling rivalry, 114–116, 123
organic compounds, 57
orgasm, female, 140
ornithology, 169
Orwell, George, 191
Ötzi (the Iceman), 181–182
overpopulation, 185

paleoanthropology, 151, 184
Paley, William, 27
Panglossian view, 34, 144
panhuman traits, 84–85
parent–offspring conflict, 111–124
Parsons, Talcott, 111
participatory anthropic principle, 48
Pascal, Blaise, 28, 103–104
Pasternak, Boris, 75
Pasteurellis pestis, 174
Pasteur Institute, 64
Pasteur, Louis, 10–11
Patchen, Kenneth, 191
Paul, (Saint), 8
peace, 146–161
penis envy, 4
Penrose, Roger, 43
Pepperberg, Irene, 100
personhood, 75
phenotype, 73
physics, 7, 11, 42–47, 88, 191
Pinker, Steven, 150
Piss-Poor Paradigms Past, 9
pitch, 55–56
The Plague (Camus), 30–31
Planck, Max, 5, 88
plant neurobiology, 102
plate tectonics, 20
polyandry, 135, 137–141, 144
polygamy, 135
polygyny, 135–138, 142–144
Pope, Alexander, 16–17, 21–22
Popper, Karl, 89
popular culture, 112
population growth, 185
Porphyry, 95

Porter, Cole, 164
power, 179–189
pregnancy, 123
primatology, 151
Prisoner's Dilemma, 165
Prometheus, 19
Proust, Marcel, 84
psychiatry, 11
psychology, 109, 122, 148
Ptolemy, 3, 22
puddle theory, 39
purpose, 107–108
Pythagoras, 104, 106
Python, Monty, 189

quantum mechanics, 20
Qur'an, 23, 53

Raleigh, Walter, 55–56
Randall, Lisa, 49
rationality, 104–106, 109
rattlesnakes, 186–187
reason, 107–108
reciprocal altruism, 164–166
reciprocity, 164–170
recurrent laryngeal nerves, 37
Rees, Martin, 45, 47
regression, 120
rejuvenation therapy, 64
relativistic mechanics, 20
religion, 28–29, 87
Religious Doctrine, 86
reproduction, 112–113, 175–176
reproductive fitness, 163–164
Research Institute of Medical
    Primatology, 63–64
responsibility, 189
reverence for life, 52–53
Ricard, Mathieu, 85–86
Rico (border collie), 99–100
right to life, 67
risk aversion, 106–107
Roberts, Alice, 8
Roman Catholic Church, 23
Romanticism, 7
Rousseau, Jean-Jacques, 104
Royal Society of London, 33
Roy, Arundhati, 189

Sagan, Carl, 8, 49, 78, 87, 91
Sagan, Lynn (Margulis), 78
The Salmon of Doubt (Adams), 39
sanity, 106
Sartre, Jean-Paul, 12, 159
Savage, Dan, 140–141
savagery, 186
Savannah baboons, 151
Schaller, George, 119
schizophrenia, 11
Schweitzer, Albert, 52–53, 61
Science, (journal), 98–100
science fiction, 59
scientific discovery, 88
scientific inquiry, 6, 9, 11–12, 87–88, 102,
     190–193
Second Law of Thermodynamics, 5, 54, 78
Selbstuberwindung (self-overcoming), 76
selection
     cosmological, 48–49
     group, 153–154, 167–170
     kin, 164, 169, 176–178
     natural, 20–21, 34, 48–49, 72–73, 113,
          153, 162–163, 169, 174
self, 68–81
self-control, 172–178
self-deception, 132
selfies, 6
self-interest and selfishness, 6, 120, 154, 156,
     164, 176–178
self-overcoming *(Selbstuberwindung),* 76
self-sacrifice, 167–168
self-sufficiency, 68–69, 115–116
Semang people, 150
Semmelweis, Ignaz, 11
Seneca, Lucius Annaeus, 54
Sepkoski, Jack, 93–94
sexual behavior, 121–122
sexual bimaturism, 136–137
sexual dimorphism, 135–136
sexual jealousy, 140
sexual reproduction, 112–113
Shakespeare, William, 86–87, 103,
     111, 129
Shanks, Niall, 45, 47
Sherman, Alan, 74
Sherrington, Charles, 77

sibling rivalry, 114–116, 123
The Silence of Animals (Gray), 147
Simon, Herbert, 107
Singer, Isaac Bashevis, 173n
Singer, Peter, 102
Sisyphus, 30–31
Skinner, B. F., 92, 157
slave morality, 168
The Sleep of Reason Produces Monsters
     (Goya), 109
Smith, John Maynard, 164
Smolin, Lee, 49
socialization, 111
social science, 112
sociobiology, 148
sociosexual competition, 136–137
Socrates, 8
Söderberg, Patrik, 150
The Song of Myself (Whitman), 80
Soviet biology, 64
Spiderman, 189
Spinoza, Baruch, 6, 43–44
spontaneous generation, 10
Squarepants, SpongeBob, 178
Stein, Gertrude, 93
stem cells, 65–66
Stevens, Hiram, 94
Stewart, Potter, 126
Stoermer, Eugene, 17–18
stotting, 133
Streep, Meryl, 141
strong force, 42
sugar fondness, 182–183
Sukhumi Primate Research Institute, 64
superego, 121
supernormal releasers, 97
sweets, 182–183
Swift, Jonathan, 4, 15–18, 103, 174
symbol behavior, 14

tapeworms, 173
tardigrades, 58–61
Tate, Nahum, 86
taxonomy, 94–95
technology, 179–189, 191
teleology, 43
Tennyson, Alfred, 192

terminology, 65
theism, 20
theology, 86
Theory of Everything, 44
theory of mind, 131
thermodynamics, 5, 54, 78
thermophiles, 56, 58
Thich Nhat Hanh, 69–70, 75–76
Thomas, Lewis, 69
Thompson, Francis, 71
thought, 93–110
The Three Body Problem (Liu), 59–60
Thucydides, 2
time, 41
Tinbergen, Niko, 96–97, 126
Tiwi people, 150
tools, 100
Trappist-1, 42
trematode worms, 174
Trisolarans, 59–60
Trivers, Robert L., 112–113, 116, 120, 165
Truman, Harry, 155
Trump, Donald, 76, 190–191
truth(s), 5, 87, 126, 190–191
truth-telling, 130–133
Tversky, Amos, 106–107
Twain, Mark, 163, 170
type societies, 150
Tyson, Neil de Grasse, 79

Underground Man, 104–106
Universe, 44–45, 79
University of British Columbia, 149

Valery, Paul, 89
vampire bats, 166
van Osten, William, 101
Vesalius, 90
Vibrio cholerae, 175
violence, 142, 146–161, 185
vitalism, 10, 56–58
Voltaire, 144
von Frisch, Karl, 126
von Tschermak, Erich, 88
Voronov, S. A., 64

Waiting for Godot (Beckett), 29–30
Wallace, Alfred Russell, 41, 88
Waorani people, 148
war and warfare, 146–161, 186
Warren, Robin, 11–12
Warsaw, Poland, 14–15
Washington, George, 10
Wason test, 108
Watt, James, 18
weak force, 42
weaning conflict, 114–115
weapons, 94, 185–186, 188
Weber, Max, 3
Wegener, Alfred, 20
Weinberg, Steven, 90
"Western, Educated, Industrialized,
    Rich, and Democratic" (WEIRD)
    research, 149
West, Rebecca, 106
Wheeler, John, 48
Whiggery, 89
Whig history, 89
White, Ellen, 23
White, Leslie, 14
Whitman, Walt, 80
Wiener, Norbert, 73
will, 172–178
Wilson, Edward O., 151, 188
Winnicott, D. W., 111
wisdom, 6, 8–9
Wittgenstein, Ludwig, 49
Wöhler, Friedrich, 57
Wolf Man, 122–123
wolves, 70–71
World War II, 109

Yanomamo people, 147–149, 155
Yellowstone National Park, 70–71, 119
"You Gotta Have Skin" (Sherman), 74
Yugoslavia, 106

Zahavi, Amotz, 133
Zappa, Frank, 173
zeedonks, 64
zoology, 64, 132–133